T0155839

SpringerBriefs in Mathematics

SpringerBriefs in Mathematics showcases expositions in all areas of mathematics and applied mathematics. Manuscripts presenting new results or a single new result in a classical field, new field, or an emerging topic, applications, or bridges between new results and already published works, are encouraged. The series is intended for mathematicians and applied mathematicians.

More information about this series at http://www.springer.com/series/10030

SBMAC SpringerBriefs

The **SBMAC SpringerBriefs** series publishes relevant contributions in the fields of applied and computational mathematics, mathematics, scientific computing, and related areas. Featuring compact volumes of 50 to 125 pages, the series covers a range of content from professional to academic.

The Sociedade Brasileira de Matemática Aplicada e Computacional (Brazilian Society of Computational and Applied Mathematics, SBMAC) is a professional association focused on computational and industrial applied mathematics. The society is active in furthering the development of mathematics and its applications in scientific, technological, and industrial fields. The SBMAC has helped to develop the applications of mathematics in science, technology, and industry, to encourage the development and implementation of effective methods and mathematical techniques for the benefit of science and technology, and to promote the exchange of ideas and information between the diverse areas of application.

http://www.sbmac.org.br/

Sara Díaz Cardell • Amparo Fúster-Sabater

Cryptography with Shrinking Generators

Fundamentals and Applications of Keystream Sequence Generators Based on Irregular Decimation

 Springer

Sara Díaz Cardell
Department of Mathematics
Institute of Mathematics, Statistics and
Scientific Computing (IMECC)
University of Campinas (UNICAMP)
Campinas, Brazil

Amparo Fúster-Sabater
Institute of Physical and Information
Technologies (ITEFI)
C.S.I.C.
Madrid, Spain

ISSN 2191-8198　　　　　ISSN 2191-8201　(electronic)
SpringerBriefs in Mathematics
ISBN 978-3-030-12849-4　　　ISBN 978-3-030-12850-0　(eBook)
https://doi.org/10.1007/978-3-030-12850-0

Library of Congress Control Number: 2019933100

Mathematics Subject Classification: 94A60, 68Q80, 94A55

This Springer imprint is published by the registered company Springer Nature Switzerland AG.
The registered company address is: Gewerbestrasse 11, 6330 Cham, Switzerland

To Henrique, Aida, Jose, Leticia, Fran, Gonzalo and Jimena.

To Angel Miguel, Isabel, Miguel, Irene and Lydia.

Preface

Cryptography is a hot research area which affects all of us whether or not we are aware of it on a daily basis: connecting to the internet, unlocking a car door with a remote control device, sending or receiving WhatsApp messages or paying with a credit card. Other emerging applications, including e-health, the so-called Internet of Things (IoT) and smart buildings, are making cryptography even more ubiquitous.

There are two kinds of cryptosystems: symmetric and asymmetric. Asymmetric systems use a public key to encrypt a message and a private key to decrypt it. In symmetric systems, the same secret key is used to encrypt and decrypt a message. Furthermore, symmetric key ciphers are implemented as either block ciphers or stream ciphers (even though sponge-based constructions are also emerging). After the irruption of public-key cryptography, with its new and fascinating possibilities, it seemed that secret-key cryptography had been confined to a few irrelevant applications. Nothing is further from the truth, since there are many examples of secret-key cryptography hardware devices and stream ciphers are simple and the fastest among all encryption/decryption procedures.

The ECRYPT stream cipher project, called eSTREAM, revitalized the field of stream ciphers after the widespread deployment of AES and highlighted the importance of stream ciphers in many technological areas. It is not surprising that the conjunction of simplicity and speed in a single process preserves its leading part in any application.

This book addresses a particular class of stream ciphers known as irregular decimation-based sequence generators. This family has given rise to a wide collection of designs, high-speed implementations and the subsequent development of cryptanalytic attacks. Our main purpose is to gather in a comprehensive survey all the research literature on this topic. In particular, we shall focus on the four most important decimation-based generators: the shrinking, self-shrinking, modified and generalized self-shrinking generator. Moreover, parallel modelling in terms of linear cellular automata reveals the implicit linearity of such generators, which were paradoxically designed and conceived to exhibit non-linearity. Indeed, these results confirm the subtle thought by J. L. Massey: "Linearity is the curse of the cryptographer" (Crypto'89).

Our approach is based on traditional linear feedback shift registers (one bit per LFSR stage), meaning that the decimation criterion depends on a single binary digit. Due to the novel design of such registers over extended fields (more than one bit per LFSR stage), new decimation criteria have to be designed. As a consequence, the family of decimation-based sequence generators must be redefined according to this new scenario. Therefore, the new challenge is to design secure irregular decimation-based generators over extended fields.

We have tried to make this text self-contained and accessible to graduate or advanced undergraduate students, as well as engineers and enthusiasts. The mathematical background is not very demanding, and we clarify the main concepts with many illustrative examples, tables and figures.

The authors wish to thank Professor Carlile Lavor for the opportunity to write this book and Professor Henrique Sá Earp for many valuable corrections and suggestions to the manuscript. We also thank the Springer team for their support and encouragement, with special gratitude to our editors Vinodhini Srinivasan and Robinson dos Santos.

Campinas, Brazil   Sara Díaz Cardell
Madrid, Spain   Amparo Fúster-Sabater
February 2018

Contents

Acronyms

AES	Advanced Encryption Standard
ANF	Algebraic Normal Form
CA	Cellular automaton
CLK	Clock of the Cryptographic Scheme
CLK2	Secondary Clock of the Cryptographic Scheme
CPU	Central Processing Unit
DES	Data Encryption Standard
DLFSR	Dynamic Linear Feedback Shift Register
ECRYPT	European Network of Excellence in Cryptology
eSTREAM	European stream cipher project
FSM	Finite state machine
GB	Gigabyte
GSM	Global System for Mobile Communications
GSSG	Generalized self-shrinking generator
HW	Hardware
IV	Initialization vector
IM	Interleaved matrix
KB	Kilobyte
LC	Linear complexity
LCT	Linear Consistency Test
LFSR	Linear Feedback Shift Register
MHz	Megahertz
MSSG	Modified self-shrinking generator
NIST	National Institute of Standards and Technology
OTP	One-Time Pad
PN	Pseudonoise
RAM	Random Access Memory
SG	Shrinking generator
SSD	Solid-State Drive
SSG	Self-shrinking generator
SW	Software
XOR	Exclusive-OR Logic Operation

Chapter 1
Introduction to Stream Ciphers

The word cryptology comes from two Greek roots meaning "hidden" and "word", and is the generic name used to describe the entire field of secret communications. Cryptology clearly splits into two opposite but complementary disciplines: cryptography and cryptanalysis. Cryptography seeks methods to ensure the secrecy of a confidential message while cryptanalysis seeks to break such methods in order to recover the confidential message. In fact, the original message upon which the cryptographer applies the cryptographic transformation is called the plaintext message, or simply the *plaintext*. The result of this transformation is called the ciphertext message, or simply the *ciphertext*, or most often the *cryptogram*. In order to control the enciphering process, the cryptographer always makes use of an exclusive information, the *key*. The general assumption in cryptology is that the cryptanalyst has full access to the cryptogram. Moreover, at present the Kerckhoff's assumption [64] is almost universally adopted by the cryptological community. According to this precept, the security of the cipher must reside entirely in the key or, equivalently, the entire cryptosystem except for the value of the key is known to the cryptanalyst.

Cryptographic systems provide secrecy by means of transformations. Depending on the type of transformation and on the type of key, the cryptosystems are commonly classified into symmetric and asymmetric cryptographic systems.

In symmetric cryptography (also called secret key cryptography), there is only a single piece of private and necessarily secret information the so-called key. Such a secret key is known to and used by the sender to encrypt the original message into a ciphertext as well as such a secret key is also known to and used by the legitimate receiver to decrypt the ciphertext into the original message. It is assumed that this double operation of encryption/decryption is impossible to be carried out without the knowledge of the secret key. Thus, in symmetric cryptography the key is shared by both legitimate communicating parties. As a result, any two users who want to communicate secretly must have previously exchanged the key in a safe way, e.g., using a trusted courier. All cryptography from ancient times until 1976

S. Díaz Cardell, A. Fúster-Sabater, *Cryptography with Shrinking Generators*,
SpringerBriefs in Mathematics, https://doi.org/10.1007/978-3-030-12850-0_1

was exclusively based on symmetric methods. Nowadays symmetric cryptography is still in widespread use, particularly for data encryption and integrity check of messages.

In asymmetric cryptography (also called public key cryptography), there are two pieces of information where at least one of which is computationally infeasible to recover from the knowledge of the other. One of the pieces is the *encryption key* (public piece of information) used by the sender to encrypt the information to be secured. The other one is the *decryption key* (secret piece of information) used by the receiver to decrypt the received ciphertext. Thus, in asymmetric cryptography each legitimate communicating party has a double key: a secret key non-shared with anyone and a public key that is known to everyone simply looking up in a public directory. In 1976, public key cryptography arose as an entirely different concept in the field of cryptography. It was first introduced by W. Diffie and M. Hellman in their mythic paper "New directions in cryptography" [21]. Asymmetric ciphers are currently used in digital signatures and key establishment as well as for classical data encryption.

Conceptually speaking, asymmetric methods seem to be more adequate for cryptographic purposes as they avoid the crucial problem of key distribution. Nevertheless, due to the nature of its operations public key algorithms are much slower than secret key algorithms. In practice, an hybrid solution is required: the key exchange is performed by public key methods and then the encryption/decryption procedure is performed by secret key methods.

Traditionally, symmetric cryptography has been split into stream ciphers and block ciphers, which can be easily distinguished.

Stream ciphers encrypt bits individually. This operation is performed by adding a bit from a pseudorandom sequence (keystream sequence) to a plaintext bit. Thus, the generation of the ciphertext is reduced to an addition of bits. Stream ciphers are synchronous when the keystream sequence depends only on the secret key and are asynchronous when the keystream sequence also depends on the ciphertext. Most practical stream ciphers are synchronous as the totality of stream ciphers considered in this book are. As example of asynchronous cipher, the cipher feedback (CFB) mode can be referenced [78, Chapter 5].

Block ciphers encrypt an entire block of plaintext bits at a time by using the same secret key. Thus, the encryption of any plaintext bit inside a given block depends on every other plaintext bit in the same block. In practice, the majority of block ciphers have a block length of 128 bits such as the Advanced Encryption Standard (AES) [19]. Nevertheless, important block ciphers with a block length of 64 bits, e.g., the Data Encryption Standard (DES) [77] or the triple DES (3DES) [78, Chapter 3], can also be referenced, although they are not recommended any more for practical applications.

In addition, different designs of sponge-based constructions [2] complete the previous categorization. Indeed, a sponge function is a generalization of both hash functions, which have a fixed output length, and stream ciphers, which have a fixed input length.

Nowadays stream ciphers are the fastest and simplest among the encryption procedures so they are implemented in many technical applications, e.g., cell phones, Internet traffic or embedded devices with little computational resources. In the following sections of this chapter, main characteristics and generalities of stream ciphers will be revised. In addition, a brief description of the most important families of stream ciphers that can be found in the literature will also be provided.

1.1 Stream Cipher

The basic problem in stream cipher design is to generate from a short and truly random key a long pseudorandom bit sequence called the keystream sequence. For encryption, the sender performs the bitwise XOR (exclusive-OR) operation among the bits of the original message or plaintext and the keystream sequence. The result is the ciphertext to be sent to the receiver. For decryption, the receiver generates the same keystream sequence, performs the same bitwise XOR operation between the received ciphertext and the keystream sequence and recovers the original message. Notice that both encryption and decryption procedures use the same operation what simplifies considerably the software/hardware implementation of this type of cipher. Moreover, such an operation is nothing but the mod 2 addition or XOR logic operation, an extremely simple and balanced operation. At any rate, the security of a stream cipher depends on the nature of the keystream sequence employed.

The precursor of the modern stream cipher is the one-time pad (OTP) or Vernam cipher invented by Gilbert Vernam in 1917. According to [52] and [78, Chapter 2], Vernam built an electromechanical machine for teletypewriter communications. The plaintext was fed into the machine as one punched paper tape and the keystream sequence as the second tape of the same characteristics. This was the first time in which encryption and transmission was automated in one machine. The main features of the OTP are:

1. The keystream sequence is only known to the legitimate communicating parties.
2. The keystream sequence is generated by a true random number generator.
3. The keystream sequence needs to be as long as the plaintext.
4. Every keystream sequence is used only once.

Under the previous conditions, the OTP is unconditionally secure or, equivalently, exhibits a mathematically proven security. Condition 1 is an habitual requirement for symmetric cryptography. Concerning conditions 2, 3 and 4, the implications are much more severe. In fact, conditions 2 and 3 mean that the keystream sequence must be generated from a physical process with length at least equal to the length of the original message, then duplicated and sent to sender and receiver through a secure channel. Moreover, we need one bit of key for each bit of plaintext. Condition 4 means that the process of generation and delivery of the sequence must be repeated every time that a secure communication is required. Clearly, the OTP is an impractical cryptographic procedure for a massive use in e-mail encryption, mobile phones, smart cards, web browsers or similar daily applications even though it is unconditionally secure.

In practice, stream cipher substitutes the truly random keystream sequence for a pseudorandom keystream sequence generated from a short random key, e.g., no more than 128 bits, and a deterministic algorithm (the keystream generator) publicly known. Once sender and receiver have exchanged the random key in a safe way and generated the same keystream sequence, then the encryption/decryption procedure is performed as described in the Vernam cipher. Due to the substitution of a truly random keystream sequence (Vernam cipher) for a pseudorandom keystream sequence (stream cipher), the latter cipher procedure does not exhibit unconditional security. In practice, the best we can do is to design keystream generators assumed to be computationally secure. In terms of symmetric cryptography, it means that there is no cryptanalytic attack with a better complexity than an exhaustive search. In brief, stream cipher is just an approximation to OTP; the more the keystream sequence looks like a truly random sequence, the more secure the stream cipher will be.

Due to its conceptual simplicity, stream cipher is the fastest among the present cryptosystems so it is easy to find many of its technological applications everywhere, e.g., the algorithms A5 in GSM communications (see Sect. 1.2.6), the encryption system E0 in Bluetooth network specifications [24], the algorithm RC4 used in Microsoft Word processor and Microsoft Excel spreadsheet [80] or the SNOW 3G Generator [49] for wireless communication of high-speed data with 4G/LTE (long-term evolution) technology.

Finally, it must be stressed that stream cipher is mainly the cipher system for military and diplomatic purposes, for which this type of symmetric cryptography is well suited. This is the reason why many important designs and practical applications of stream ciphers are and will be condemned to the most absolute obscurantism.

1.1.1 A Basic Structure in Stream Cipher: The Linear Feedback Shift Register (LFSR)

In this subsection, we provide some basic notation and concepts that will be used throughout the book.

Let p be a prime, m a positive integer and $q = p^m$. Let $\mathbb{F}_q$ denote a finite field with q elements. The order of an element $\alpha \in \mathbb{F}_q$, denoted by $\text{ord}(\alpha)$, is the smallest positive integer k such that $\alpha^k = 1$. An element α with order $q - 1$ is called a primitive element in $\mathbb{F}_q$. The primitive elements are exactly the generators of $\mathbb{F}_q^*$, the multiplicative group consisting of the nonzero elements of $\mathbb{F}_q$. Thus, a finite field $\mathbb{F}_q$ consists of 0 and appropriate powers of a primitive element.

Let $\{a_i\}$, $i = 0, 1, 2, \ldots$, be a sequence over $\mathbb{F}_p$ if $a_i \in \mathbb{F}_p$, for all $i \geq 0$. The sequence $\{a_i\}$ is periodic if and only if there exists an integer $T > 0$ such that $a_{i+T} = a_i$ holds for all $i \geq 0$.

Let L be a positive integer, and let $c_0, c_1, \ldots, c_{L-1}$ be given elements of the finite field $\mathbb{F}_p$. A sequence $\{a_i\}$ of elements of $\mathbb{F}_p$ satisfying the relation

$$a_{i+L} = c_1 a_{i+L-1} + c_2 a_{i+L-2} + \ldots + c_{L-1} a_{i+1} + c_L a_i, \qquad i \geq 0, \qquad (1.1)$$

is called an Lth order linear recurring sequence in $\mathbb{F}_p$. The terms $a_0, a_1, \ldots, a_{L-1}$, which determine uniquely the rest of the sequence, are referred to as the initial values. A relation of the form given in (1.1) is called an Lth order homogeneous linear recurrence relationship. The monic polynomial of degree L

$$p(x) = x^L + c_1 x^{L-1} + c_2 x^{L-2} + \ldots + c_{L-1} x + c_L \in \mathbb{F}_p[x] \qquad (1.2)$$

is called the characteristic polynomial of the linear recurring sequence and the sequence $\{a_i\}$ is said to be generated by $p(x)$. The polynomial of the lowest degree in the set of characteristic polynomials of $\{a_i\}$ over $\mathbb{F}_p$ is called the minimal polynomial of $\{a_i\}$ over $\mathbb{F}_p$. For a survey of linear recurring sequences over finite fields, the interested reader is referred to [59].

In this book, we will consider sequences defined exclusively over the binary field $\mathbb{F}_2$, i.e., $p = 2$ and $q = 2^m$, while the extension field will be denoted by $\mathbb{F}_{2^m}$. It should be noticed that the analysis provided here can be extended to sequences over any prime extension $\mathbb{F}_{p^m}$.

The generation of linear recurring sequences can be implemented on linear feedback shift registers (LFSRs). These devices handle information in the form of elements of $\mathbb{F}_2$ and they are based on shifts and linear feedback. A conventional or Fibonacci LFSR consists of L *interconnected stages* numbered $0, 1, \cdots, L - 1$ (from left to right) capable of storing one bit, the *feedback* or *connection polynomial*[1] and the *initial state* (stage contents at the initial instant). In addition, a clock controls the movement (shifts) of data. During each unit of time, the following operations are performed (see Fig. 1.1):

1. The content of stage 0 is output and forms part of the output sequence.
2. The content of stage n is moved to stage $n - 1$ for each n ($1 \leq n \leq L - 1$).
3. The new content of stage numbered $L - 1$ is the feedback bit calculated by adding mod 2 the previous contents of a fixed subset of stages (taps) determined by the feedback polynomial.

In terms of practical implementation, the Galois LFSRs appear as alternative structures that generate exactly the same linear recurring sequences as those of Fibonacci LFSRs. More precisely, in Galois LFSRs the taps are not concatenated so they can be updated in parallel, increasing the speed of execution.

For a minimal polynomial $p(x)$ as this one defined in Eq. (1.2), the output of the LFSR with nonzero initial state is the string of elements $\{a_0, a_1, a_2, a_3, \ldots\}$

---

[1]The feedback polynomial of the LFSR and the minimal polynomial of its linear recurrence relationship are reciprocal polynomials.

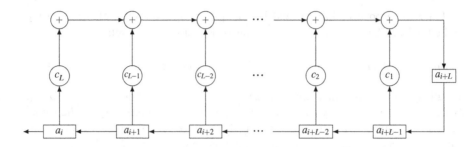

Fig. 1.1 LFSR of length L

generated in intervals of one-time unit (see Fig. 1.1). If the minimal polynomial of the linear recurring sequence is primitive [7], then the LFSR is called maximal-length LFSR and its output sequence has period $2^L - 1$, see [41]. This output sequence is called **PN-sequence** (pseudonoise sequence) or ***m*-sequence** (maximal sequence). In the sequel, all LFSRs considered will be maximal-length LFSRs. In the cryptographic literature, the LFSR minimal polynomial is simply termed as characteristic polynomial.

Linear Feedback Shift Registers are used in many of the keystream generators that have been proposed in the literature. The main reasons for such a continuous use can be enumerated as follows:

1. LFSRs provide high performance when used as sequence generators.
2. They are particularly well-suited to hardware implementations.
3. They generate output sequences with large period and good statistical properties. In fact, such sequences satisfy Golomb's pseudorandomness postulates [41].
4. Due to their simple structure, LFSRs can be readily analysed by means of algebraic techniques.

According to Golomb's pseudorandomness postulates [41], the PN-sequences are balanced (the difference between the number of ones and zeros in one period of the sequence does not exceed one), the number of binary runs (consecutive ones or consecutive zeros) occurs with the right probability (half of runs have length one, one-fourth length two, one-eighth length three, etc., as long as for each of these lengths the number of one-runs equals the number of zero-runs) and their autocorrelation function is two-valued.

At first glance, sequences obtained from maximal-length LFSRs might look like good candidates to keystream sequences. Nevertheless, as explained later, they do not satisfy a fundamental condition required to all cryptographic sequence and related with the linear character of these registers.

The linear complexity (LC) of a sequence $\{a_i\}$ is defined as the length of the shortest LFSR that can generate such a sequence or, equivalently, the order of the shortest linear recurrence relationship satisfied by such a sequence. In a general sense, linear complexity is related with the amount of sequence that is needed to determine the whole sequence. The Berlekamp–Massey algorithm efficiently computes the length and characteristic polynomial of the shortest LFSR given at least $2 \cdot LC$ sequence bits, see [63]. Indeed, the running time of the Berlekamp–Massey algorithm is $O(N^2)$, where N is the length of the sequence under consideration.

Linear complexity is a much used metric of the security of a keystream sequence. In cryptographic terms, linear complexity must be as large as possible. The recommended value is approximately half the sequence period, $LC \simeq T/2$. According to the own definition of linear complexity, sequences generated from maximal-length LFSRs of length L will have a LC of value equal to L, what is too far from the recommended value of $T/2 \simeq 2^{L-1}$. Consequently, LFSRs should never be used alone as keystream generators. Indeed, the linear complexity of their output sequences has to be increased before such sequences are employed for cryptographic purposes.

1.2 LFSR-Based Sequence Generators

In order to overcome the low LC inherent to the sequences generated by LFSRs, in the literature several approaches are proposed. In the sequel, different methods of designing keystream sequence generators will be briefly described. All of them pursue the same goals:

- To preserve the good statistical properties of the PN-sequences.
- To increase the LC of the sequences generated by LFSRs.

Besides LC, other properties must be taken into account when keystream sequences are considered.

In fact, balancedness is one of the good statistical properties that every keystream sequence must satisfy. Roughly speaking, a binary sequence is balanced if it has approximately the same number of ones as zeros. Due to the long period of a keystream sequence ($T \simeq 10^{38}$ bits in current cryptographic applications), it is not feasible to produce an entire cycle of such a sequence and then count the number of ones and zeros. Therefore, in practice, portions of the keystream sequence are chosen randomly and the frequency test (monobit test) [70, Chapter 5] is applied to all these subsequences. If all of them pass the statistical test, then the sequence is accepted as being balanced. Nevertheless, passing the frequency test merely provides probabilistic evidence that the generator produces a balanced sequence. In the literature, balancedness of keystream sequences has been treated in a deterministic way [31, 32, 45]. Indeed, there are simple binary models based on the sequence generator parameters that allow one to compute the exact number of ones in the keystream sequence without producing the whole sequence. The same can be applied to the computation of the number of runs of any length in a keystream sequence [25].

In brief, long period, balancedness, good run distribution and large linear complexity are some necessary (never sufficient) conditions for a keystream sequence to be considered secure [32]. In addition, such sequences have to pass a battery of tests (NIST tests [76], DIEHARD tests [61] and Tuftests [62]) to be accepted as cryptographic sequences. Traditionally, the key of these stream cipher cryptosystems is the initial contents of the LFSRs included in the design. Next, a quick overview of the main families of LFSR-based sequence generators is introduced.

1.2.1   Non-linear Combination Generators

A classical technique for destroying the linearity inherent to LFSRs is to use N LFSRs working synchronously. The keystream sequence $\{s_j\}$ is produced as the image of a non-linear Boolean function f whose N variables at time t are the corresponding output bits of the N registers [59]. The function f is expressed in algebraic normal form (ANF) as the mod 2 addition (XOR logic operation) of distinct nth order products in its N variables with $0 \leq n \leq N$. The non-linear order of f is the maximum order of the terms appearing in its ANF. This construction is illustrated in Fig. 1.2, where $s(t) = s_t$ is the tth term of the keystream sequence. These keystream sequence generators are called non-linear combination generators (or non-linear combiners) and f is the combining function.

The security of those generators is conditioned by the properties of such a function. In general, the non-linear combination generators provide sequences with large period, good statistical properties and moderate linear complexity. Depending on the combining function choice, these generators can be vulnerable to certain cryptanalytic attacks (e.g., correlation attacks).

As a representative example of this type of generator, we can analyse the well-known Geffe generator [70, Chapter 6], see Fig. 1.3.

This generator is made up of three maximal-length LFSRs of lengths L_1, L_2, L_3, which are pairwise relatively prime. The combining function is

$$f(x_1, x_2, x_3) = x_1 x_2 \oplus x_2 x_3 \oplus x_3,$$

where the symbol $\oplus$ means the XOR logic operation. $LFSR_2$ acts as selector switching the output between $LFSR_1$ and $LFSR_3$. The keystream sequence $\{s_j\}$ obtained from the Geffe generator has period $T = (2^{L_1} - 1)(2^{L_2} - 1)(2^{L_3} - 1)$ and

Fig. 1.2 Non-linear combiner

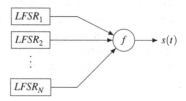

Fig. 1.3 Geffe generator

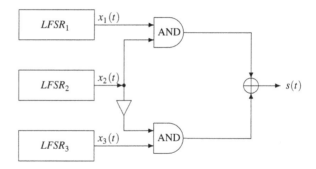

Table 1.1 Truth table for the Geffe generator

$x_1(t)$	$x_2(t)$	$x_3(t)$	$s(t)$
0	0	0	0
0	0	1	0
0	1	0	0
0	1	1	1
1	0	0	1
1	0	1	0
1	1	0	1
1	1	1	1

linear complexity $LC = L_1L_2 + L_2L_3 + L_3$. Recall that in this type of combination generators and under a variety of conditions [89] the linear complexity of the output sequence satisfies $LC = f(L_1, L_2, L_3)$. Thus, the LC of the output sequence is closely related to the order of the combining function.

Concerning balancedness, we can see in Table 1.1 that the combining function f is balanced as well as the three PN-sequences generated by the LFSRs are. In [31], a general expression in terms of L_i $(i = 1, 2, 3)$ provides the exact number of ones in the output sequence of a Geffe generator. Such an expression is

$$No(1's) = 2^{L_1-1}2^{L_2-1}(2^{L_3} - 1) + (2^{L_1} - 1)(2^{L_2} - 1)2^{L_3-1}.$$

For lengths of the LFSRs in a cryptographic range $L_i \simeq 60$, the number of ones in the output sequence is $No(1's) \simeq T/2$. Consequently, the generated sequence can be considered as a quasi-balanced sequence.

The Geffe generator is cryptographically weak because information about the successive bits from $LFSR_1$ and $LFSR_3$ leaks into the output sequence. In fact, let $x_1(t), x_2(t), x_3(t)$ be the tth output bit of $LFSR_i$ $(i = 1, 2, 3)$ and $s(t) = s_t$ the tth output bit of the keystream sequence, respectively. According to Table 1.1, the correlation probability between $x_1(t)$ and $s(t)$ and between $x_3(t)$ and $s(t)$ is

$$P(s(t) = x_1(t)) = P(s(t) = x_3(t)) = \frac{3}{4},$$

although the correlation probability between $x_2(t)$ and $s(t)$ is given by $P(s(t) = x_2(t)) = \frac{1}{2}$. Consequently, the Geffe generator is vulnerable to a simple correlation attack as it is shown in [70, Chapter 6].

In general, the combining function $f(x_1, x_2, \cdots, x_N)$ must be carefully selected in order to avoid a statistical dependence between any subset of the N PN-sequences and the keystream sequence. This condition can be guaranteed if f is chosen to be mth order correlation immune [70, Chapter 6], m being an integer $m < N$. The non-linear order of a Boolean function and the correlation immunity are properties closely related in the sense that if $f(x_1, x_2, \cdots, x_N)$ is chosen to be mth order correlation immune, then its non-linear order is at most $N - m$.

Different principles of design for good non-linear combination generators based on binary LFSR structures can be recommended:

1. Use maximal-length LFSRs to get long period and good short-term statistics in the output sequence.
2. Choose the LFSR lengths $L_1, L_2, \cdots, L_N$ to be relatively prime, i.e., $\gcd(L_i, L_j) = 1$ for $i \neq j$, to get long period.
3. Apply the practical design of balanced sequence combination generators given in [31] to get a balanced or quasi-balanced output sequence.
4. Choose the non-linear order of f to obtain a good compromise between linear complexity and correlation immunity.
5. Choose the non-linear function f to have terms of each order to get good confusion.
6. Let the key determine some terms of the function f.

The Geffe generator is an example of memoryless combination generator. Nevertheless, with the use of memory the combining function f becomes a non-linear finite state machine (FSM) which greatly increases the number of options available for these structures [88, Chapter 9]. In this case, the memoryless combining function is responsible for the level of correlation immunity and the balanced distribution of the output, whereas the next-state function is responsible for the level of non-linearity. The summation generator is a good example of memory combination generator where memory is included in the carry bit [88, Chapter 9]. Moreover, integer addition is a cryptographically useful function as it is extremely non-linear when viewed over the binary field $\mathbb{F}_2$.

1.2.2   Non-linear Filters

Another general technique for destroying the linearity inherent to LFSRs is to use a non-linear filter. In this case, the keystream sequence $\{s_j\}$ is generated as the image of a non-linear Boolean function f in the L stages of a unique LFSR, that is, the L variables of the Boolean function are the binary contents of the LFSR stages at each time instant t. This construction is illustrated in Fig. 1.4. These keystream sequence generators are called non-linear filters and f is the filtering function. Period and

Fig. 1.4 Non-linear filter

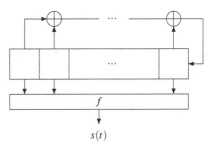

statistical properties of the filtered sequences are characteristics deeply studied in the literature, see references [70, 78, 88].

Concerning the linear complexity, it can be stated that if the non-linear order of the Boolean function is k, then the linear complexity of the filtered sequence is at most

$$LC_{max} = \sum_{i=1}^{k} \binom{L}{i}.$$

Nevertheless, the problem of determining the exact value of the linear complexity attained by filtering functions is still an open problem [27, 55, 60]. At any rate, several contributions to the linear complexity of non-linearly filtered sequences can be quoted:

1. In [88, Chapter 5], Rueppel proves that the output sequence from non-linear filters including a unique term of equidistant stages has a linear complexity lower bounded by $LC \geq \binom{L}{k}$, where L is the LFSR length and $k \approx L/2$ the order of the filtering function. For (L, k) in a cryptographic range, e.g., (128, 64), the lower bound is quite large.
2. Later, in [79] the equivalence between the root presence test [88, Chapter 5] and the discrete Fourier transform approach is established, which allows the author to give lower bounds on the linear complexity for new classes of filtering functions.
3. In [56], the authors provide an improved lower bound $LC \geq \binom{L}{k} + \binom{L}{k-1}$ on the linear complexity of filtered sequences. In any case, this lower bound is only applicable to non-linear filters of order $k \in [2, 3, L - 1, L]$, which is outside the standard cryptographic range.
4. Finally, in [26] a method of computing all the non-linear filters applied to an LFSR with $LC \geq \binom{L}{k}$ is developed. The procedure is based on the concept of equivalence classes of non-linear filters and is performed by means of additions and shiftings of filtering functions coming out from different classes. The method formally completes the family of non-linear filters found in the literature with a large guaranteed linear complexity.

Concerning balancedness, it can be proved that if the filtering function f is a balanced function, then the filtered sequence will have the same period as that of the underlying LFSR [91, Theorem 1]. In addition, a binary model to compute the exact number of ones in the output sequence of a non-linear filter can be found in [32, subsection 3.2]. The computational method analyses the form of the Boolean function f and is based exclusively on the handling of binary strings by means of logic operations. The proposed model serves as a deterministic alternative to existing probabilistic methods for checking balancedness in this type of sequence generators.

Different principles of design for a good non-linear filter based on a binary LFSR structure can be recommended:

1. Use a maximal-length LFSR to get long period and good short-term statistics in the output sequence.
2. Choose a non-linear order k in the filtering function f to get large linear complexity, e.g., $k \approx L/2$, where L is the LFSR length.
3. Include a linear term and several terms of each small order in f to get good short-term statistics.
4. Apply the computational method given in [32] to check balancedness in the output sequence.
5. Include some terms of every order up to k in f to get good confusion.
6. Let the key determine some terms of the function f.

As a representative example of this type of generator, we can describe the Hitag2 generator. Hitag2 is an encryption algorithm designed by NXP Semiconductors that is used in electronic vehicle immobilizers and anti-theft devices [98]. Hitag2 uses a proprietary stream cipher with a key of 48 bits. Such a generator is a non-linear filter made up of a 48-stage LFSR and a filtering function. The feedback polynomial includes the binary contents of 16 stages in the feedback loop. The filtering function consists of three different functions f_a, f_b and f_c, see Fig. 1.5. In fact, f_a and f_b take as their four input variables the contents of different LFSR stages, while f_c takes as its five input variables the output bits of the functions f_a and f_b. Next,

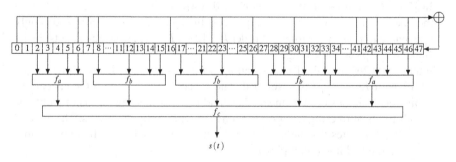

Fig. 1.5 Hitag 2

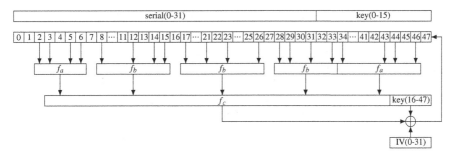

Fig. 1.6 Hitag 2 initialization

the f_c output variable is the corresponding bit $s(t)$ of the keystream generator. The previous functions are defined as follows:

$$f_a(i) = (0x2C79)_i,$$
$$f_b(i) = (0x6671)_i,$$
$$f_c(i) = (0x7907287B)_i,$$

where the output of these functions for the input i is the ith bit of the above hexadecimal values.

Previously to the keystream generation, Hitag2 needs an initialization phase to fill the 48 stages of the LFSR. The initialization procedure is described as follows. In addition to the 48-bit key, this sequence generator uses a 32-bit serial number and a 32-bit initialization vector (IV). In fact, the LFSR is filled with the 32 bits of the serial number and the first 16 bits of the key, see Fig. 1.6. Next, the cipher works in an autonomous mode for 32 cycles where the LFSR feedback bit is the result of the mod 2 addition among the corresponding key bit (16–47), the corresponding IV bit (0–31) and the Hitag2 output bit. Once the 32 cycles have been performed, the LFSR stage contents are the LFSR initial state for the register to start generating the keystream sequence.

Due to its short key, Hitag2 is considered an insecure stream cipher. Different algebraic attacks have been proposed in the literature, e.g., algebraic attacks [18, 93], attacks with a specific hardware [92, 99] or an exhaustive search attack with low cost technology [34]. Due to cost reasons, the automotive industry is surprisingly reluctant to migrate to other more secure products with a longer key.

1.2.3 Clock-Controlled Generators

In Sect. 1.2.1, the N LFSRs of a non-linear combination generator were clocked regularly, that is, the shift of data in all the registers was controlled by the same clock. Nevertheless, the main idea behind a clock-controlled generator is that the

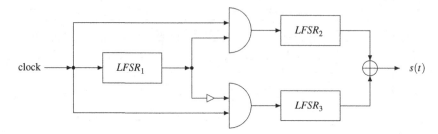

Fig. 1.7 Alternating-step generator

output sequence of one LFSR controls the clock of at least other LFSR. This irregular clocking of any LFSR is a simple strategy that introduces non-linearity into the output sequence.

As the most representative example of this type of generator, the alternating-step generator [43] is described in Fig. 1.7. The alternating-step generator uses three maximal-length LFSRs, notated $LFSR_i$ $(i = 1, 2, 3)$, of lengths L_1, L_2, L_3, which are pairwise relatively prime. In order to generate the output sequence $\{s_j\}$ the following steps are repeated:

1. The register $LFSR_1$ is clocked.
2. If the output bit of $LFSR_1$ equals 1, then $LFSR_2$ is clocked while $LFSR_3$ is not clocked but repeats its previous output bit.
3. If the output bit of $LFSR_1$ equals 0, then $LFSR_3$ is clocked while $LFSR_2$ is not clocked but repeats its previous output bit.
4. The tth bit of the keystream sequence $s(t)$ is the mod 2 addition between the output bits of $LFSR_2$ and $LFSR_3$ at the time instant t.

For the first clock cycle, the previous output bit of registers $LFSR_2$ and $LFSR_3$ is taken to be 0. The alternating-step generator is based on the stop-and-go generator of Beth and Piper [3] where only one of the LFSR was irregularly clocked.

The keystream sequence $\{s_j\}$ obtained from the alternating-step generator has period $T = 2^{L_1}(2^{L_2} - 1)(2^{L_3} - 1)$ and its linear complexity LC satisfies the inequality

$$(L_1 + L_3)\, 2^{L_1-1} < LC \le (L_2 + L_3)\, 2^{L_1}.$$

The distribution of patterns in the output sequence is almost uniform. In fact, if $\mathbf{S}_s$ denotes a pattern of any s consecutive bits, then the probability P that $\mathbf{S}_s$ appears in the output sequence is given by $P(\mathbf{S}_s) \simeq (\frac{1}{2})^s$.

Recall that in this type of clock-controlled generators the linear complexity of the output sequence is lower bounded by 2^{L_1-1}, that is, LC is exponential in the length of one of the LFSRs. It means that the fact of introducing irregular clocking makes increase dramatically the value of the linear complexity.

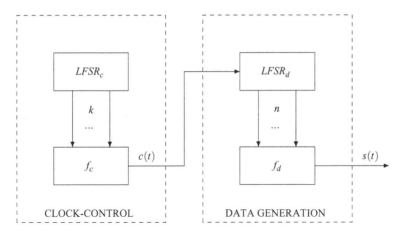

Fig. 1.8 LILI keystream generators

The security of the alternating-step generator is based on a right choice of the lengths L_i that should be about the same, that is, $L_1 \simeq L$, $L_2 \simeq L$ and $L_3 \simeq L$. In that case, the best known attack on this generator is a divide and conquer attack on the control register $LFSR_1$ [43] that takes approximately 2^L steps. Thus, if $L = 128$, then the generator is secure against this type of attack. Certain correlation attacks against clock-controlled shift registers can also be found in [35, 39] with approximately the same computational complexity as that one of the attack previously mentioned.

Among other interesting clock-controlled keystream generators, we can refer:

1. The Gollmann cascade generator [70, Chapter 6] made up of a succession of m maximal-length LFSRs of the same length L. The clock of the $LFSR_i$ is controlled by all the previous $LFSR_j$ with $j < i$. The output sequence exhibits large period $T = (2^L - 1)^m$ and excellent $LC \geq L\,(2^L - 1)^{m-1}$.
2. The LILI family of keystream generators [91] that can be viewed as a clock-controlled non-linear filter, see Fig. 1.8. The clock-control block ($LFSR_c$ + non-linear filter f_c) determines the shift of the $LFSR_d$ to whom stages a non-linear filter f_d is applied. This type of design offers large period and LC. However, some algebraic attacks can be found in the literature [16, 17]. At any rate, an attack against LILI-128 [17] can take 2^{57} CPU clocks but the requirements of intercepted bits are far from being practical.

1.2.4 Decimation-Based Generators

The underlying idea of this type of generators is the irregular decimation of a PN-sequence according to the bits of another one. The result of this decimation is an output sequence that will be used as keystream sequence in the cryptographic procedure of encryption/decryption.

Irregularly decimated generators produce good cryptographic sequences characterized by long periods, good correlation, excellent run distribution, balancedness, simplicity of implementation, etc. Inside the family of irregularly decimated generators, we can enumerate: (a) the shrinking generator proposed by Coppersmith, Krawczyk and Mansour [15] that includes two LFSRs, (b) the self-shrinking generator designed by Meier and Staffelbach [67] involving only one LFSR, (c) the generalized self-shrinking generator or family of generators proposed by Hu and Xiao [46] that includes the self-shrinking generator and (d) the modified self-shrinking generator introduced by Kanso [53] that is related with the family of generalized self-shrinking generators. Indeed, the generalized self-shrinking generator can be seen as a specialization of the shrinking generator as well as a generalization of the self-shrinking generator. In fact, the output sequence of the self-shrinking generator is just an element of the family of generalized self-shrinking sequences.

This book focuses on decimation-based sequence generators with application in stream ciphers. Next chapters address systematically diverse features of these generators and their corresponding keystream sequences.

1.2.5 Dynamic LFSR Generators

In [74], Mita et al. proposed a new keystream sequence generator for cryptographic application based on LFSRs that they called "topology with dynamic linear feedback shift register" (DLFSR). In fact, such a topology consists in changing dynamically the feedback polynomial of the main LFSR included in the design. In this way, the output sequence of this type of generator $\{s_j\}$ is nothing but the concatenation of different portions of distinct PN-sequences. This new topology was first introduced in a generic way by means of one LFSR whose feedback polynomial was updated according to the stage contents of a secondary LFSR. In this proposal, the authors provided only series of experimental data from this particular implementation. Later in [82], Peinado et al. analysed and modelled different cryptographic parameters of the generated sequences, e.g., period, linear complexity, autocorrelation, run distribution, etc.

Basically, a DLFSR consists of:

1. A main $LFSR$ with n stages and N_p primitive feedback polynomials that will be successively applied according to a particular order determined by the feedback module.
2. A feedback control module including, among other structures, a secondary $LFSR$ with m stages and a unique primitive feedback polynomial. This module is going to control the feedback polynomial of the main register.

Although the method of generating output sequence is common to all DLFSRs, such generators can be classified into different categories depending on the operation mode:

1. DLFSR generators that apply the different N_p primitive feedback polynomials in the same order to generate the same number of output bits with each applied polynomial [68, 69, 83].
2. DLFSR generators that apply the N_p feedback polynomials in the same order to produce a different number of output bits with each one of the applied polynomials [84].
3. The most general case in which the DLFSR generators apply the N_p feedback polynomials in a pseudorandom order to produce with each polynomial a different number of output bits [1, 14, 54].

As illustrative example of DLFSR generators, Fig. 1.9 depicts a generic DLFSR generator belonging to the third category above mentioned. In fact, it is a generalization of the DLFSR module designed in [84]. The proposal represented in Fig. 1.9 is made up of two LFSRs (main and secondary registers) with n and m stages, respectively, and a counter that counts backwards from a particular value determined by the state of the secondary LFSR. At the same time, the counter controls CLK2 the clock of the secondary LFSR. The choice of the feedback polynomial applied to the main LFSR is determined by k_1 bits of the secondary LFSR among the N_p primitive feedback polynomials previously selected. Both LFSRs are initialized with their

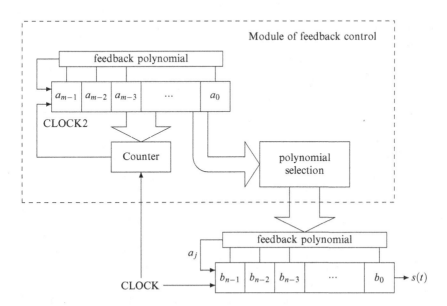

Fig. 1.9 DLFSR

corresponding initial states (key of the keystream generator). After the initialization process, the generation of the output sequence is detailed as follows:

1. The counter is initialized with k_2 bits of the secondary LFSR with $k_2 \leq \log_2 m$.
2. The main LFSR starts generating bits of the output sequence.
3. Simultaneously, the counter starts counting backwards until the value 0 is obtained. At that moment, the clock CLK2 is activated and the secondary LFSR generates a bit.
4. The new secondary LFSR state determines by means of k_1 bits the new feedback polynomial as well as by means of k_2 bits the new value of the counter.

The design here presented improves the period and linear complexity of the output sequence when compared with the same parameters obtained in DLFSR proposals [14, 74].

1.2.6 Other Types of Keystream Sequence Generators

Other types of keystream generators not included in the previous subsections can be also described. In this subsection, we consider the multiple speed Massey–Rueppel generator and the algorithms A5/1 and A5/2 used in GSM (global system for mobile communications) technology.

The Massey–Rueppel generator [65] is a keystream sequence generator employing multiple speed LFSRs. Therefore, the speed factor is treated as an additional variable in the sequence generation. The underlying idea in multiple speed generators is that, when a speed factor is introduced, a single LFSR with a fixed feedback polynomial can generate the PN-sequence corresponding to other LFSR with different feedback polynomial. Thus, multiple speed gives a new dimension to the design of secure generators.

In Fig. 1.10, an example of the simplest Massey–Rueppel generator is depicted. It consists of only two maximal-length LFSRs, notated $LFSR_i$ $(i = 1, 2)$, of lengths L_1 and L_2, respectively, which are relatively prime. The lower register $LFSR_2$ is clocked at a clock rate greater than that of the upper register $LFSR_1$. The $LFSR_2$ clock rate, notated d, is the speed factor of the generator and can be kept secret as a part of the key. The output bit $s(t)$ is the mod 2 addition of the logic products (AND operation) among the contents of the corresponding stages in both registers. The output sequence exhibits an excellent short-term statistics, no leakage and a long period of value $T = (2^{L_1} - 1)(2^{L_2} - 1)$. The weakness of this generator is its moderate linear complexity as LC is proportional to the lengths of both LFSRs. In fact, $LC = L_1 L_2$ although the scheme can be iterated to N LFSRs to get greater linear complexity of value $LC = L_1 L_2, \ldots, L_N$. The fact of changing the speed factors allows the user to generate distinct output sequences keeping unchanged the LFSRs included in the design.

Next, a different family of keystream generators is also described. The A5 stream cipher was designed to protect the over-the-air privacy of GSM telephone

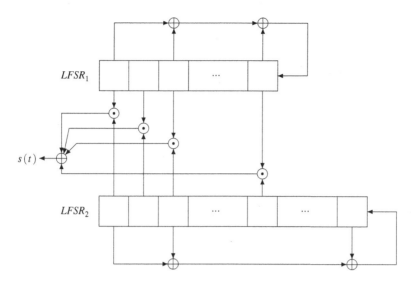

Fig. 1.10 Massey–Rueppel generator

conversations. This algorithm has two main variants: the stronger A5/1 version used by millions of customers in Europe and the weaker A5/2 version used by another millions of customers in other markets. The functional schemes of both versions will never be published. At any rate, they were reverse engineered by M. Briceno and later confirmed against official test vectors [6].

A GSM conversation is sent as a sequence of frames where every frame contains 228 bits. Each GSM conversation is encrypted by a session key K derived from algorithm A8 included in the more general algorithm COMP128, see [4]. For each frame, the key K is mixed with the corresponding frame counter (a known number of 22 bits) and the result serves as initial state of the LFSRs. From this initial state, the keystream generator first produces 100 bits that will be rejected and then the corresponding 228 keystream bits. Such bits are mod 2 added with the 228 bits of the conversation frame in order to produce the 228 bits of the ciphered conversation frame. The same process is repeated systematically for each one of the successive frames. Recall that for each conversation frame the key K is always the same only the frame counter is different.

In the following, the description of both generators A5/1 and A5/2 and their cryptanalysis are detailed.

The A5/1 generator is made up of three maximal-length LFSRs, notated $LFSR_i$ ($i = 1, 2, 3$), of lengths $L_1 = 19$, $L_2 = 22$ and $L_3 = 23$. According to Fig. 1.11, the taps of $LFSR_1$ are at bit positions 14, 17, 18 and 19 (numbered from right to left); the taps of $LFSR_2$ are at bit positions 21 and 22; and the taps of $LFSR_3$ are at bit positions 8, 21, 22 and 23. The internal state of A5/1 at time t

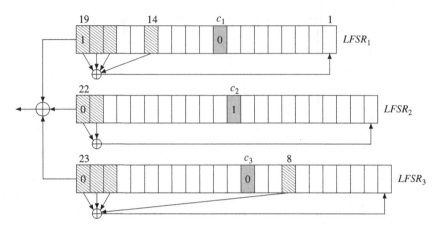

Fig. 1.11 Algoritmo A5/1

is the binary contents of the LFSRs at this particular moment. The three LFSRs are clocked in a stop/go fashion using the following majority rule:

1. Each LFSR has a single clocking stage, notated c_1, c_2 and c_3, corresponding to bit 9 in $LFSR_1$, bit 11 in $LFSR_2$ and bit 11 in $LFSR_3$.
2. At each clock cycle, the majority function F defined as

$$F(c_1, c_2, c_3) = c_1 c_2 \oplus c_1 c_3 \oplus c_2 c_3$$

   is computed.
3. Only those LFSRs whose clocking stages agree with the majority function are actually clocked.
4. At each clock cycle, one output bit is produced as the mod 2 addition of the most significant bits in the three LFSRs.

Recall that at each clock cycle either two or three LFSRs are clocked. Moreover each LFSR moves with probability $3/4$ and stops with probability $1/4$.

Different cryptanalyses of the A5/1 generator have appeared in the literature. Particularly important is the work developed in [37], where the author describes a general time-memory trade-off attack concluding that it is possible to find the A5/1 key. This attack is based on the knowledge of a certain amount of intercepted keystream sequence and a precomputed table storing internal states and their corresponding output sequence portions. Comparing the intercepted sequence with these output prefixes, an intermediate state in some frame could be identified. Then, A5/1 runs backwards until getting the initial state of this particular frame. The key can be extracted from any frame initial state reversing the effect of the known frame counter. At any rate, for this cryptanalytic attack the requirements of intercepted keystream sequence and space in the table were unrealistic.

Nevertheless, keeping in mind all these ideas but in a more refined way, Biryukov et al. succeeded in performing an outstanding cryptanalysis [4] that revealed the insecurity of the A5/1 generator. In fact, they introduced the concept of "special states" in A5/1 or states able to produce output bits starting with a particular pattern *alpha* of length $k = 16$. The idea was scanning the intercepted sequence until such a particular pattern was encountered. Once an intermediate state had been identified, the rest of the cryptanalytic attack was the same as that one described by Golic [37]. Indeed, the fact of using just special states allowed the cryptanalysts to reduce dramatically the amount of intercepted sequence and size of the precomputed table until rather realistic levels. In brief, the easy generation of the special states, the frequent A5/1 setup routine where the same key is repeated in each frame initialization and the high probability of getting a coincidence between intercepted and stored patterns (guaranteed by the birthday paradox) are the most remarkable weaknesses of this keystream generator.

The A5/2 generator is a modified version of the previous generator including a fourth LFSR. Figure 1.12 depicts the general scheme of the A5/2 algorithm. In fact, the registers $LFSR_i$ ($i = 1, 2, 3$) are the same as those ones used in the A5/1 generator, while the additional register $LFSR_4$ is an LFSR of length $L_4 = 17$ whose taps are at bit positions 12 and 17. Notice that $LFSR_4$ controls the shift of the remaining registers: each $LFSR_i$ is clocked when its corresponding clocking variable c_i equals 1. According to Fig. 1.12, the majority function F defined as before appears four times in different parts of the scheme. At each clock cycle, one output bit is produced as the mod 2 addition of the most significant bits in the three LFSRs plus the results of three majority functions taking values in the stages of $LFSR_i$ ($i = 1, 2, 3$).

As most representative cryptanalysis of the A5/2 generator, the algebraic attack described in [85] can be referenced. It consists in writing out a system of equations that relates the state variables of the $LFSR_i$ ($i = 1, 2, 3$) with the output bit by

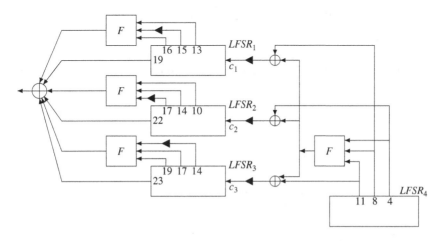

Fig. 1.12 Algorithm A5/2

means of a clock-control sequence produced by the register $LFSR_4$ starting at a particular initial state. In the worst case, all the $2^{17} - 1$ possible initial states of $LFSR_4$, paradoxically the shortest register, must be considered. Each one gives rise to a different system of equations. The linearization of these equations is performed by substitution of the non-linear terms by new and linear variables. After having written about 620 equations in each one of those linearized systems, many linearly dependent equations appear. The knowledge of four frames of keystream sequence and the linear dependences allow the cryptanalysts to reconstruct the following bits. Then, comparing reconstructed sequences with the intercepted sequence, the right keystream sequence is selected. The time complexity of this attack is proportional to 2^{17}. Frequent reinitialization of the frames, small number of skipped bits (just 100 bits rejected) in the initialization process and/or a bad distribution of taps in the LFSRs seem to be the origin of these linear dependences that this attack successfully exploits.

1.3   eSTREAM

The eSTREAM, the European stream cipher project, was a multi-year effort launched by the ECRYPT (European Network of Excellence in Cryptology) in November 2004. The goal of eSTREAM was to "promote the design of efficient and compact stream ciphers suitable for widespread adoption". After public discussions at the State of the Art of Stream Ciphers (SASC) a workshop held in Bruges (October 2004), ECRYPT published its call for stream cipher primitives and the result was the eSTREAM project [96].

The stream cipher proposals were classified into two different profiles:

1. Profile 1. Stream ciphers for software applications with high throughput requirements.
2. Profile 2. Stream ciphers for hardware applications with restricted resources such as limited storage, gate count, or power consumption.

After the call for primitives, 34 candidates were submitted to eSTREAM. Apart from a panel of experts, it was the cryptographic community at large who, after a formal evaluation in three phases, selected the most suitable stream cipher proposals for the final portfolio.

The main evaluation criteria were:

1. Security.
2. Performance when compared to AES in some appropriate mode (e.g., counter mode).
3. Performance when compared to other submissions.
4. Justification and supporting analysis.
5. Simplicity and flexibility.
6. Completeness and clarity of the submission.

Moreover, several requirements concerning the length of the key and the initialization vector (IV) as well as further technical characteristics were also defined.

The required parameter values were:

1. Profile 1.
   A key length of 128 bits must be accommodated.
   An IV length of at least one of 64 or 128 bits must be accommodated.
2. Profile 2.
   A key length of 80 bits must be accommodated.
   An IV length of at least one of 32 or 64 bits must be accommodated.

Software performance is an aspect particularly significant for Profile 1 candidates. In fact, software performance can be measured in many different ways. In order to make comparisons as fair as possible, eSTREAM developed a testing framework to assure that all stream cipher proposals were submitted to the same tests under the same circumstances. This testing framework and documentation available in [94] has been and continues to be used by other researchers outside of eSTREAM.

As a result of the project, a portfolio of eight new and promising stream ciphers was announced in April 2008, see Table 1.2. Later the eSTREAM portfolio was revised in September 2008 when M. Hell and T. Johansson [44] published a cryptanalytic attack in real time against the F-FCSR-H v2. As a consequence, the ECRYPT had to eliminate the cipher F-FCSR-H v2 from the previous list. At the present moment, the eSTREAM portfolio contains the ciphers listed in Table 1.3 with seven stream ciphers.

Descriptions, possible weaknesses and reports on software/hardware performance of not only the proposals in the portfolio but also of all eSTREAM candidates can be found in the corresponding links of the official eSTREAM website [94]. The portfolio is periodically revised as the algorithms mature. Different reviews of the eSTREAM portfolio were published in October 2009, January 2012 and another one recently, see [94]. At the same time, a volume published by Springer [87] in 2008 provides full specifications of all 16 ciphers that reached the final phase of the

Table 1.2 First eSTREAM portfolio

Profile 1 (SW)	Profile 2 (HW)
HC-128	Grain v1
Rabbit	MICKEY v2
Salsa20/12	Trivium
SOSEMANUK	F-FCSR-H v2

Table 1.3 Final eSTREAM portfolio

Profile 1 (SW)	Profile 2 (HW)
HC-128	Grain v1
Rabbit	MICKEY v2
Salsa20/12	Trivium
SOSEMANUK	

eSTREAM project. In fact, it is a very detailed survey covering both the software- and the hardware-oriented finalists. In addition, a prototype of an ASIC containing all Profile 2 candidates was designed and fabricated on $0.18\,\mu m$ CMOS, as part of the eSCARGOT project [95].

Keeping in mind that the goal of eSTREAM was to stimulate works in the area of stream ciphers, undoubtedly the project has been a great success. It served to:

1. revitalize the field of stream ciphers after the widespread deployment of AES, and
2. identify two areas where a dedicated stream cipher design might offer advantages over block ciphers:

   - areas where exceptionally high throughput is required in software, and
   - areas where exceptionally low resource consumption is required in hardware.

Over the following years the eSTREAM proposals have been assessed with regard to both security and practicality by the cryptographic community, and the results presented at major conferences and specialized workshops dedicated to the state of the art of stream ciphers. Until now, the stream ciphers included in the portfolio list remain unchanged.

Chapter 2
Keystream Generators Based on Irregular Decimation

In this chapter, we study the definition and the principal characteristics of the main keystream generators based on irregular decimation: the shrinking generator, the self-shrinking generator, the modified self-shrinking generator and the generalized self-shrinking generator.

First of all, we need to recall the concept of decimation. Let $\{v_i\}$, $i = 0, 1, 2, \ldots$, be a linear recursive sequence over a finite field. The decimation of this sequence by distance d is a new sequence $\{v_{d \cdot i}\}$, $i = 0, 1, 2, \ldots$, obtained by taking every dth term of $\{v_i\}$ (see [22]).

Example 2.1 Consider the LFSR of length 3 with characteristic polynomial $p(x) = 1 + x + x^3$. If we consider the initial state $\{1\ 0\ 0\}$, the PN-sequence generated is the following:

$$\{a_i\} = \{1\ 0\ 0\ 1\ 0\ 1\ 1\ \ldots\}.$$

Since $p(x)$ is primitive, $\{a_i\}$ has maximum-period equal to 7. Now, if we denote by $\{b_i\}$ the sequence obtained decimating $\{a_i\}$ by distance 2:

$$
\begin{array}{cccccc}
b_0 & b_4 & b_2 & b_5 & b_3 & b_6 \\
\uparrow & \uparrow & \uparrow & \uparrow & \uparrow & \uparrow \\
1 & 0 & 0 & 1 & 0 & 1\ 1\ldots
\end{array}
$$

this sequence has the form $\{b_i\} = \{a_{2i}\} = \{1\ 0\ 0\ 1\ 0\ 1\ 1\ \ldots\}$. Notice that $\{b_i\}$ is the same PN-sequence $\{a_i\}$. This is due to the fact that the period of the PN-sequence and the distance of decimation are relatively primes, that is, $\gcd(7, 2) = 1$ [41]. ∎

2.1 Shrinking Generator

In this section we present the main characteristics of the first generator based on irregular decimation, the shrinking generator.

2.1.1 Definition and Basic Features

The **shrinking generator** (SG) was introduced by Coppersmith, Krawczyk and Mansour in [15]. This generator was very attractive in that moment, due to its conceptual simplicity, since it combines two binary maximum-length LFSRs in a simple way. The output sequence of the generator is produced by shrinking the output sequence of one LFSR under the control of the other. In other words, the PN-sequence $\{a_i\}$, $i = 0, 1, 2, \ldots$, produced by one of the registers, denoted by R_1, decimates the PN-sequence $\{b_i\}$, $i = 0, 1, 2, \ldots$, produced by the other register, denoted by R_2. Let L_1 and L_2, with $\gcd(L_1, L_2) = 1$, be the number of stages (or length) of R_1 and R_2 and $p_1(x), p_2(x) \in \mathbb{F}_2[x]$ their characteristic polynomials, respectively. We consider these polynomials to be primitive, to assure the output sequences are maximum-period or PN-sequences. We will denote by $\{s_j\}$, $j = 0, 1, 2, \ldots$, the output sequence of the generator and we will call it the **shrunken sequence**. The decimation rule is very simple:

$$\begin{cases} \text{If } a_i = 1, \text{ then } s_j = b_i. \\ \text{If } a_i = 0, \text{ then } b_i \text{ is discarded,} \end{cases}$$

that is, the output bit of R_2 is taken if the current bit of R_1 is 1, otherwise it is discarded.

The key of the generator is the initial states of both registers and the characteristic polynomials, which are recommended to be part of the key.

When $\gcd(L_1, L_2) = 1$, the period of the shrunken sequence is

$$T = 2^{L_1-1} \left(2^{L_2} - 1\right),$$

and its linear complexity, denoted by LC, satisfies $L_2 2^{L_1-2} < LC \leq L_2 2^{L_1-1}$. Furthermore, the shrunken sequence is balanced and has other good cryptographic properties [15]. Therefore, this scheme is supposed to be suitable for practical implementation in encryption procedures.

Let us see an illustrative example of this generator.

Example 2.2 Consider R_1 the LFSR with characteristic polynomial $p_1(x) = 1 + x + x^2$ and initial state $\{1\ 0\}$. The PN-sequence generated by R_1, in this case, is $\{a_i\} = \{1\ 0\ 1\ \ldots\}$. Consider also R_2 the LFSR with characteristic polynomial $p_2(x) = 1 + x + x^3$ and initial state $\{1\ 0\ 0\}$. The PN-sequence produced is $\{b_i\} =$

{1 0 0 1 0 1 1 ...}. Then, the shrunken sequence can be computed in the following way:

$$\{a_i\} : 1\ 0\ 1\ 1\ 0\ 1\ 1\ 0\ 1\ 1\ 0\ 1\ 1\ 0\ 1\ 1\ 0\ 1\ ...$$
$$\{b_i\} : 1\ \cancel{0}\ 0\ 1\ \cancel{0}\ 1\ 1\ \cancel{0}\ 0\ \cancel{0}\ 0\ 1\ \cancel{0}\ 1\ 0\ \cancel{0}\ 1\ 0\ \cancel{0}\ 1\ ...$$
$$\{s_j\} : 1\ \ \ 0\ 1\ \ \ 1\ 1\ \ \ 0\ 0\ \ \ 0\ 1\ \ \ 1\ 0\ \ \ 1\ 0\ \ \ 1\ ...$$

The shrunken sequence $\{s_j\}$ has period 14 and, thanks to the Berlekamp–Massey algorithm [63], it is not difficult to check that its characteristic polynomial is $p(x)^2 = (1 + x^2 + x^3)^2$, consequently its linear complexity equals 6. ∎

Despite its simplicity, there are currently no known attacks better than exhaustive search of the initial states of the registers, when the characteristic polynomials are secret.

It is worth noticing that there may be multiple initial states that produce the same keystream sequence (equivalent keys). For example, let us consider the registers used in Example 2.2. If we consider initial states {0 1} and {0 1 1}, respectively, we obtain the following shrunken sequence:

$$\{a_i\} : 0\ 1\ 1\ 0\ 1\ 1\ 0\ 1\ 1\ 0\ 1\ 1\ 0\ 1\ 1\ 0\ 1\ 1\ 0\ 1\ 1\ ...$$
$$\{b_i\} : \cancel{0}\ 1\ 1\ \cancel{0}\ 0\ 0\ \cancel{0}\ 0\ 1\ \cancel{0}\ 1\ 0\ \cancel{0}\ 1\ 0\ \cancel{0}\ 1\ 1\ \cancel{0}\ 0\ 1\ ...$$
$$\{s_j\} : \ \ 1\ 1\ \ \ 0\ 0\ \ \ 0\ 1\ \ \ 1\ 0\ \ \ 1\ 0\ \ \ 1\ 1\ \ \ 0\ 1\ ...$$

On the other hand, if we consider initial states {1 1} and {1 1 1}, respectively, we obtain the following shrunken sequence:

$$\{a_i\} : 1\ 1\ 0\ 1\ 1\ 0\ 1\ 1\ 0\ 1\ 1\ 0\ 1\ 1\ 0\ 1\ 1\ 0\ 1\ 1\ 0\ ...$$
$$\{b_i\} : 1\ 1\ \cancel{0}\ 0\ 0\ \cancel{0}\ 0\ 1\ \cancel{0}\ 1\ 0\ \cancel{0}\ 1\ 0\ \cancel{0}\ 1\ 1\ \cancel{0}\ 0\ 1\ \cancel{0}\ ...$$
$$\{s_j\} : 1\ 1\ \ \ 0\ 0\ \ \ 0\ 1\ \ \ 1\ 0\ \ \ 1\ 0\ \ \ 1\ 1\ \ \ 0\ 1\ \ \ ...$$

which is the same as before. Due to leading 0s in the first two PN-sequences, both keys generate the same shrunken sequence. For this reason, we always consider initial states that start with 1. Thus, the effective key size is smaller than the key space.

From now on, we consider two registers R_1 and R_2, with primitive characteristic polynomials $p_1(x)$, $p_2(x) \in \mathbb{F}_2[x]$, lengths L_1 and L_2 and $\gcd(L_1, L_2) = 1$, respectively. Besides, the PN-sequences generated by both registers are denoted by $\{a_i\}$ and $\{b_i\}$ and have periods $T_1 = 2^{L_1} - 1$ and $T_2 = 2^{L_2} - 1$, respectively. We assume without loss of generality that $a_0 = 1$.

2.1.2 Characteristic Polynomial and Interleaved PN-Sequences

In this section, we will see that the shrunken sequence is constructed interleaving one unique PN-sequence and the form of its characteristic polynomial.

Theorem 2.1 ([9]) *The 2^{L_1-1} sequences obtained decimating the shrunken sequence by distance 2^{L_1-1} starting in positions $0, 1, 2, \ldots, 2^{L_1-1}-1$, respectively, are PN-sequences with characteristic polynomial*

$$p(x) = \left(x + \alpha^{T_1}\right)\left(x + \alpha^{2T_1}\right)\left(x + \alpha^{4T_1}\right)\cdots\left(x + \alpha^{2^{L_2-1}T_1}\right),$$

where $\alpha \in \mathbb{F}_2^{L_2}$ is a root of the polynomial $p_2(x)$ and $T_1 = 2^{L_1} - 1$ is the period of the PN-sequence generated by R_1.

All the interleaved PN-sequences of the shrunken sequence are generated by the same characteristic polynomial, this means that all of them are shifted versions of the same PN-sequence.

It is worth remarking that, since α is a primitive element of the field $\mathbb{F}_2^{L_2}$, $p_2(x)$ needs to be primitive.

Example 2.3 Consider two registers, R_1 and R_2, with characteristic polynomials $p_1(x) = 1 + x + x^3$ and $p_2(x) = 1 + x + x^4$ and initial states $\{1\,0\,0\}$ and $\{1\,0\,0\,0\}$, respectively. Denote by $\{a_i\}$ and $\{b_i\}$ the PN-sequences generated by R_1 and R_2, respectively. The shrunken sequence generated by these registers has period $T = 60$ and is given by

$$\{s_j\} = \{1000111110100001100101101100110101000010111000110111101011011\ldots\}.$$

If we decimate the shrunken sequence $\{s_j\}$ by distance $2^{L_1-1} = 4$ starting in positions $0, 1, 2$ and 3, respectively, we obtain four interleaved PN-sequences:

$$
\begin{aligned}
\{s_{4j}\}: &\quad 1 \quad 1\,1 \langle 0\rangle 1\, \boxed{0}\, 1\,1\,0\, \textcircled{0}\, 1\,0\,0\,0\,1 \ldots \\
\{s_{4j+1}\}: &\quad \textcircled{0}\ 1\,0\ \ 0\ \ 0\,1\,1\,1\,1\ 0\ 1\,0\,1\,1\,0 \ldots \\
\{s_{4j+2}\}: &\quad \boxed{0}\ \ 1\,1\ \ 0\ \ 0\,1\,0\,0\,0\ 1\ 1\,1\,1\,0\,1 \ldots \\
\{s_{4j+3}\}: &\quad \langle 0\rangle 1\,0\ \ 1\ \ 1\,0\,0\,1\,0\ 0\ 0\,1\,1\,1\,1 \ldots
\end{aligned}
\tag{2.1}
$$

According to Theorem 2.1, the characteristic polynomial of the interleaved PN-sequences is given by

$$p(x) = \left(x + \alpha^7\right)\left(x + \alpha^{14}\right)\left(x + \alpha^{28}\right)\left(x + \alpha^{56}\right) = 1 + x^3 + x^4,$$

where $\alpha \in \mathbb{F}_{2^4}$ is a root of $p_2(x)$. Therefore, the four interleaved PN-sequences are shifted versions of the same PN-sequence generated by $p(x)$. In expression (2.1), we can check that the bits $\textcircled{0}$, $\boxed{0}$ and $\langle 0\rangle$ in the PN-sequence $\{s_{4j}\}$ represent the starting points of the sequences $\{s_{4j+i}\}$ $(1 \leq i \leq 3)$, respectively. ∎

Corollary 2.1 ([9, Corollary 1]) *If $L_2 = L_1 + 1$, then the polynomial $p(x)$, given in Theorem 2.1, is the reciprocal polynomial of $p_2(x)$.*

It is worth reminding that the reciprocal polynomial of the polynomial $r(x) = r_0 + r_1 x + \cdots + r_{l-1} x^{l-1} + r_l x^l$ is of the form $r^*(x) = r_l + r_{l-1} x + \cdots + r_1 x^{l-1} + r_0 x^l$.

Now, we introduce the form of the characteristic polynomial of the shrunken sequence.

Theorem 2.2 ([28]) *The characteristic polynomial of the shrunken sequence has the form* $p(x)^m$, *for* $2^{L_1-2} < m \leq 2^{L_1-1}$ *with* $p(x)$ *as in Theorem 2.1.*

Notice that $p(x)^{2^{L_1-1}}$ always generates the shrunken sequence, but sometimes this polynomial might not be the characteristic polynomial of lowest degree. For instance, consider again the shrunken sequence generated in Example 2.3. In this case, we computed the polynomial $p(x) = 1 + x^3 + x^4$. Now, we know that $p(x)^4 = \left(1 + x^3 + x^4\right)^4$ generates the shrunken sequence and, since $p(x)^3$ does not generate it, we can assume that $p(x)^4$ is its characteristic polynomial.

Example 2.4 Consider the registers with characteristic polynomials $p_1(x) = 1 + x^2 + x^5$ and $p_2(x) = 1 + x + x^2 + x^3 + x^4 + x^5 + x^7$, respectively. Consider the shrunken sequence generated by these two registers, which has period $T = 2^4(2^7 - 1) = 2032$. According to Theorem 2.1, the polynomial $p(x)$ can be computed as:

$$p(x) = \left(x + \alpha^{31}\right)\left(x + \alpha^{62}\right)\left(x + \alpha^{124}\right)\left(x + \alpha^{248}\right)\left(x + \alpha^{496}\right)$$
$$\left(x + \alpha^{992}\right)\left(x + \alpha^{1984}\right)$$
$$= 1 + x + x^3 + x^6 + x^7,$$

where $\alpha \in \mathbb{F}_{2^7}$ is a root of $p_2(x)$. In this case, we know that $p(x)^{16}$ generates the shrunken sequence. However, it is easy to check that $p(x)^{15}$ is the characteristic polynomial. ∎

Interestingly, $p(x)$ only depends on $p_2(x)$ and L_1. This means that if we fix a primitive polynomial $p_2(x)$ and we consider any primitive polynomial with degree L_1 we always obtain the same $p(x)$.

2.1.3 Shrunken Sequences and Difference Equations

In this section, we show that the shrunken sequence is a solution of a difference equation.

The characteristic polynomial $p(x)$ (with degree L) of an arbitrary sequence $\{a_i\}$ specifies its linear recurrence relationship. This means that the element a_i can be written as a linear combination of the previous elements:

$$a_i \oplus \sum_{j=1}^{L} c_j a_{i-j} = 0, \quad i \geq L.$$

The linear recursion can be expressed as a linear difference equation:

$$\left[E^L \oplus \sum_{j=1}^{L} c_j E^{L-j} \right] a_i = 0, \quad i \geq 0, \tag{2.2}$$

with E being the one-sided shift operator that acts on the sequence terms:

$$\begin{aligned} E a_i &= a_{i+1}, \\ E^k a_i &= a_{i+k}. \end{aligned} \tag{2.3}$$

If the characteristic polynomial $p(x)$ is primitive and α is one of its roots, then $\alpha, \alpha^2, \alpha^{2^2}, \ldots, \alpha^{2^{L-1}}$ are the L different roots of such a polynomial as well as primitive elements of $\mathbb{F}_{2^L}$ [59]. Now, if the characteristic polynomial of an arbitrary sequence $\{s_j\}$ is of the form $p(x)^m$, then its roots will be the same as those of $p(x)$ but each one with multiplicity m. The corresponding difference equation is given by

$$\left[E^L \oplus \sum_{k=1}^{L} E^{L-k} \right]^m s_j = 0,$$

and its solutions are of the form $s_j = \sum_{i=0}^{L-1} \sum_{k=0}^{m-1} \binom{j}{k} A_k^{2^i} \alpha^{2^i j}$, where A_k is an arbitrary element in $\mathbb{F}_{2^L}$. Different choices of A_k give rise to different sequences $\{s_j\}$. A particular choice of A_k provides the shrunken sequence generated by $p(x)^m$.

2.1.4 Obtaining the Second PN-Sequence from the Shrunken Sequence

Given the shrunken sequence $\{s_j\}$ generated by two registers, R_1 and R_2, it is possible to compute the PN-sequences generated by both registers. In this section, we explain how to obtain the PN-sequence $\{b_i\}$ produced by R_2.

Proposition 2.1 ([11, Proposition 1]) *Let $\delta \in \{1, 2, 3, \ldots, T_2 - 1\}$ be such that $T_1 \delta = 1 \bmod T_2$. If the first PN-interleaved sequence is decimated by distance δ, then the resultant sequence is $\{b_i\}$.*

Example 2.5 Consider again the shrunken sequence obtained in Example 2.3 and consider the interleaved PN-sequences given in expression (2.1). Since $L_1 = 3$ and $L_2 = 4$, the unique value for δ such that $7\delta = 1 \bmod 15$ is $\delta = 13$. This means that if we decimate the first interleaved sequence $\{s_{4j}\}$ by distance 13, according to

Proposition 2.1, we obtain $\{b_i\}$, the PN-sequence generated by $p_2(x) = 1 + x + x^4$:

$$
\begin{array}{c}
b_0\ b_7\ b_{14}\ b_6\ b_{13}\ b_5\ b_{12}\ b_4\ b_{11}\ b_3\ b_{10}\ b_2\ b_9\ b_1\ b_8 \\
\uparrow\ \uparrow\ \ \uparrow\ \ \ \uparrow\ \ \ \uparrow\ \ \ \uparrow\ \ \ \uparrow\ \ \ \uparrow\ \ \ \uparrow\ \ \ \uparrow\ \ \ \uparrow\ \ \ \uparrow\ \uparrow\ \uparrow\ \uparrow
\end{array}
$$

$\{s_{4j}\}: \quad 1\ 1\ 1\ 0\ 1\ 0\ 1\ 1\ 0\ 0\ 1\ 0\ 0\ 0\ 1$

$\{s_{4j+1}\}: 0\ 1\ 0\ 0\ 0\ 1\ 1\ 1\ 1\ 0\ 1\ 0\ 1\ 1\ 0$

$\{s_{4j+2}\}: 0\ 1\ 1\ 0\ 0\ 1\ 0\ 0\ 0\ 1\ 1\ 1\ 1\ 0\ 1$

$\{s_{4j+3}\}: 0\ 1\ 0\ 1\ 1\ 0\ 0\ 1\ 0\ 0\ 0\ 1\ 1\ 1\ 1$

In this case $\{b_i\} = \{1\ 0\ 0\ 0\ 1\ 0\ 0\ 1\ 1\ 0\ 1\ 0\ 1\ 1\ 1\ \ldots\}$. ■

The previous proposition leads us to the following two results.

Corollary 2.2 ([11, Corollary 2]) *If the polynomials $p_1(x)$, $p_2(x) \in \mathbb{F}_2[x]$ have degrees L_1 and $L_1 + 1$, respectively, then $\delta = T_2 - 2$.*

In Example 2.5, we had that $L_2 = L_1 + 1 = 4$, then it was not necessary to solve the equation given in Proposition 2.1, it was enough to compute $\delta = T_2 - 2 = 13$.

Theorem 2.3 ([11, Corollary 1]) *If the shrunken sequence is decimated by distance $2^{L_1-1}\delta$, then the obtained sequence is the PN-sequence $\{b_i\}$.*

Example 2.6 Consider again the shrunken sequence obtained in Example 2.2. In this example we had that $L_1 = 2$ and $L_2 = 3$, then according to Corollary 2.2, $\delta = 5$. Now, according to Theorem 2.3, we know that if we decimate the shrunken sequence by distance 10:

$$
\begin{array}{c}
b_0\quad b_3\quad b_6\quad b_2\quad b_5\quad b_1\quad b_4 \\
\uparrow\quad\ \uparrow\quad\ \uparrow\quad\ \uparrow\quad\ \uparrow\quad\ \uparrow\quad\ \uparrow \\
1\ 0\ 1\ 1\ 1\ 0\ 0\ 0\ 1\ 1\ 0\ 1\ 0\ 1\ \ldots
\end{array}
$$

then we obtain again the PN-sequence generated by the second register, R_2:

$$\{b_i\} = \{1\ 0\ 0\ 1\ 0\ 1\ 1\ \ldots\}.$$

■

2.1.5 Obtaining the First PN-Sequence from the Shrunken Sequence

In this section, we analyse how to recover the PN-sequence $\{a_i\}$ produced by R_1 from the shrunken sequence $\{s_j\}$.

Assume the first interleaved PN-sequence of $\{s_j\}$ is denoted by $\{v_i\}$. Since the other interleaved sequences are shifted versions of the same PN-sequence, it means they are shifted versions of $\{v_i\}$. Then, we assume they have the form $\{v_{d_1+i}\}, \{v_{d_2+i}\}, \ldots, \{v_{d_{2^{L_1-1}-1}+i}\}$, for some positions $d_i \in \{0, 1, 2, \ldots, 2^{L_2-2}\}$:

$$
\begin{array}{llllll}
\{v_i\}: & \{v_0 & v_1 & v_2 & \cdots & v_{T_2-1} & \cdots\} \\
\{v_{d_1+i}\}: & \{v_{d_1} & v_{d_1+1} & v_{d_1+2} & \cdots & v_{d_1+T_2-1} & \cdots\} \\
\{v_{d_2+i}\}: & \{v_{d_2} & v_{d_2+1} & v_{d_2+2} & \cdots & v_{d_2+T_2-1} & \cdots\} \\
\vdots & \vdots & \vdots & \vdots & & \vdots \\
\{v_{d_{2^{L_1-1}-1}+i}\}: & \{v_{d_{2^{L_1-1}-1}} & v_{d_{2^{L_1-1}-1}+1} & v_{d_{2^{L_1-1}-1}+2} & \cdots & v_{d_{2^{L_1-1}-1}+T_2-1} & \cdots\}.
\end{array}
$$

In order to illustrate this idea, consider again Example 2.3. We had four interleaved PN-sequences that correspond to:

$$\{s_{4j}\} = \{v_i\}, \quad \{s_{4j+1}\} = \{v_{d_1+i}\}, \quad \{s_{4j+2}\} = \{v_{d_2+i}\} \quad \text{and} \quad \{s_{4j+3}\} = \{v_{d_3+i}\}.$$

In this case, the positions are $d_1 = 9$, $d_2 = 5$ and $d_3 = 3$:

$$
\begin{array}{l}
\qquad\qquad\qquad\qquad d_3=3 \quad d_2=5 \qquad d_1=9 \\
\qquad\qquad\qquad\qquad \uparrow \qquad\; \uparrow \qquad\qquad \uparrow \\
\{s_{4j}\}: \quad 1\; \; 1\,1\; \langle\Diamond\rangle\; 1\; \boxed{0}\; 1\,1\,0\; \textcircled{0}\; 1\,0\,0\,0\,1 \\
\{s_{4j+1}\}: \textcircled{0}\; 1\,0\;\; 0\; 0\,1\,1\,1\,1\; 0\; 1\,0\,1\,1\,0 \\
\{s_{4j+2}\}: \boxed{0}\;\; 1\,1\;\; 0\; 0\,1\,0\,0\,0\;\; 1\; 1\,1\,1\,0\,1 \\
\{s_{4j+3}\}: \langle\Diamond\rangle\, 1\,0\;\; 1\;\; 1\,0\,0\,1\,0\; 0\; 0\,1\,1\,1\,1
\end{array}
\tag{2.4}
$$

Before introducing the next result, it is worth reminding that a maximum-length LFSR of L stages produces a PN-sequence with 2^{L-1} ones in its first period [41].

Example 2.7 Consider the LFSR with characteristic polynomial $p(x) = 1+x+x^4$ and initial state $\{1\;1\;1\;1\}$. The PN-sequence generated by this register is given by

$$\{1\;1\;1\;1\;0\;0\;0\;1\;0\;0\;1\;1\;0\;1\;0 \ldots\}.$$

Notice that this PN-sequence has 2^3 ones in its period, that is, in its first 15 bits. ∎

Theorem 2.4 ([11, Proposition 2]) *If* $\{0, i_1, i_2, \ldots, i_{2^{L_1-1}-1}\}$ *is the set of indices of the 1s in the PN-sequence* $\{a_i\}$ *in its first period, then* $d_k = \delta \cdot i_k \bmod (2^{L_1-1} - 1)$, *for* $k = 1, 2, \ldots, 2^{L_1-1} - 1$, *where* δ *has the form given in Proposition 2.1.*

In Example 2.3, we had four interleaved PN-sequences $\{v_i\}$, $\{v_{i+d_1}\}$, $\{v_{i+d_2}\}$ and $\{v_{i+d_3}\}$ and $\delta = 13$. We know that $d_1 = 9$, $d_2 = 5$ and $d_3 = 3$ (see expression (2.4)). Then, according to Theorem 2.4, we can compute the indices $\{0, i_1, i_2, i_3\}$ of the four 1s in the first period of $\{a_i\}$ ($i_0 = 0$, without loss of generality) solving the following system:

$$\begin{cases} 13 \cdot i_1 = 9 \bmod 15 \\ 13 \cdot i_2 = 5 \bmod 15 \\ 13 \cdot i_3 = 3 \bmod 15. \end{cases}$$

Therefore, the set of indices is given by $\{0, 3, 5, 6\}$ and then the PN-sequence produced by R_1 is given by $\{a_i\} = \{1\,0\,0\,1\,0\,1\,1\ldots\}$.

2.2 Self-Shrinking Generator

The **self-shrinking generator** (SSG) was introduced by Meier and Staffelbach in [67]. They presented a simple structure using only one maximal-length LFSR, whose output sequence $\{a_i\}$ is self-decimated. The key consists of the initial state of the register and the characteristic polynomial is again recommended as part of the key.

Let L be the length and $p(x) \in \mathbb{F}_2[x]$ the characteristic polynomial of the register. We consider again $p(x)$ primitive, to assure the output sequence has maximum-period. We will denote by $\{s_j\}$, $j = 0, 1, 2, \ldots$, the output sequence of the generator and we will call it, the **self-shrunken sequence** (SS-sequence). The decimation rule is very simple,

$$\begin{cases} \text{If } a_{2i} = 1, \text{ then } s_j = a_{2i+1}. \\ \text{If } a_{2i} = 0, \text{ then } a_{2i+1} \text{ is discarded,} \end{cases}$$

that is, pairs of bits are considered: if a pair happens to take the value 10 or 11, this pair is taken to produce the bit 0 or 1, depending on the second bit of the pair. On the other hand, if a pair happens to be 01 or 00, it will be discarded.

Example 2.8 Consider the LFSR of $L = 3$ stages with characteristic polynomial $p_1(x) = 1 + x^2 + x^3$ and initial state $\{1\,0\,0\}$. The corresponding PN-sequence is given by $\{1\,0\,0\,1\,1\,1\,0\ldots\}$. Now the self-shrunken sequence can be computed in the following way:

$$R: \underbrace{1\ \ 0}_{0}\ \ 0\ \ 1\ \ \underbrace{1\ \ 1}_{1}\ \ 0\ \ 1\ \ 0\ \ 0\ \ \underbrace{1\ \ 1}_{1}\ \ \underbrace{1\ \ 0}_{0}\ \ \cdots$$

The corresponding self-shrunken sequence is given by $\{s_j\} = \{0\,1\,1\,0\ldots\}$.   ∎

The period T of a self-shrunken sequence [67] produced by a maximal-length LFSR of L stages satisfies

$$T \geq 2^{\lfloor \frac{L}{2} \rfloor}.$$

Due to experimental observations, we claim that the period of the self-shrunken sequences is always $T = 2^{L-1}$, when $L > 3$ and $p(x)$ is primitive. However, no proof has been found yet. We encourage the reader to prove this claim.

According to Meier and Staffelbach [67] and Blackburn [5], we can say that the linear complexity satisfies

$$2^{\lfloor \frac{L}{2} \rfloor} < LC \le 2^{L-1} - (L-2).$$

Again, due to experimental observations, we claim that the lower bound for the linear complexity can be improved to: $LC > 2^{L-2}$. However, no proof has been found so far. Actually, this is also a natural consequence of $T = 2^{L-1}$. We let the reader think about this open problem.

Proposition 2.2 ([33]) *The characteristic polynomial of the self-shrunken sequences has the following form:* $p_{LC}(x) = (1+x)^{LC}$, *where LC is the linear complexity of such a sequence.*

For instance, consider Example 2.8. We had the self-shrunken sequence $\{s_j\} = \{0\ 1\ 1\ 0\ \ldots\}$ produced by $p(x) = 1 + x^2 + x^3$. It is possible to check that the self-shrunken sequence has period $T = 2^{3-1}$ and its characteristic polynomial is $p_3(x) = (1+x)^3$ (see Berlekamp–Massey algorithm [63]). Consequently, the linear complexity of $\{s_j\}$ is $LC = 3$.

2.3 Modified Self-Shrinking Generator

In [53] Kanso introduced a variant of the self-shrinking generator called the **modified self-shrinking generator** (MSSG). This generator, intended for hardware implementation, uses an extended selection rule based on the XORed value of a pair of bits in the PN-sequence. The resultant sequences are balanced and have good statistical properties.

The decimation rule is very simple and can be described as follows: given three consecutive bits $\{a_{3i}, a_{3i+1}, a_{3i+2}\}$, $i = 0, 1, 2, \ldots$, of a PN-sequence $\{a_i\}$, the output sequence $\{s_j\}$ is computed as

$$\begin{cases} \text{If } a_{3i} + a_{3i+1} = 1 \text{ then } s_j = a_{3i+2}, \\ \text{If } a_{3i} + a_{3i+1} = 0 \text{ then } a_{3i+2} \text{ is discarded.} \end{cases}$$

The output sequence $\{s_j\}$ is known as the **modified self-shrunken sequence** (MSS-sequence).

Example 2.9 Let us consider the LFSR of three stages with characteristic polynomial $q(x) = 1 + x^2 + x^3$ and initial state $\{1\ 1\ 1\}$. The PN-sequence generated by this register is given by $\{1\ 1\ 1\ 0\ 1\ 0\ 0\ \ldots\}$. In this case, the modified self-shrunken sequence can be computed as follows:

$$\{a_i\}: \underbrace{1\ 1} \cancel{\underbrace{0\ 1}} \underbrace{\textcircled{0}} \underbrace{0\ 1} \underbrace{\textcircled{1}} \underbrace{1\ 0} \underbrace{\textcircled{1}} \underbrace{0\ 0} \cancel{\underbrace{1\ 1}} \cancel{\underbrace{1\ 0}} \underbrace{\textcircled{0}} \ldots$$
$$\oplus: \quad\ 0 \qquad 1 \qquad\ 1 \qquad\ 1 \qquad\ 0 \qquad\ 0 \qquad\ 1$$

The sequence $\{s_j\} = \{0\ 1\ 1\ 0\ \ldots\}$ (encircled bits) is the MSS-sequence generated by $q(x)$. ∎

Now, we are ready to study the properties of this generator. According to [53], if we consider a maximal-length LFSR of L (odd) stages, then:

1. The period T of the MSS-sequence satisfies

$$2^{\lfloor \frac{L}{3} \rfloor} \le T \le 2^{L-1}.$$

2. The linear complexity LC of the MSS-sequence satisfies

$$2^{\lfloor \frac{L}{3} \rfloor - 1} \le T \le 2^{L-1} - (L-2).$$

Although the MSS-sequences seem to have lower bounds on the period and linear complexity than those of the SSG, Kanso claimed that these sequences provide a higher level of security against several well-known attacks. Besides, Kanso demonstrated that the MSS-sequences possess better randomness properties than those of the SSG. In next section, we will see that both sequences belong to the family of generalized self-shrunken sequences.

2.4   Generalized Self-Shrinking Generator

In [46] Hu and Xiao introduced a specialization of the shrinking generator and a generalization of the self-shrinking generator. This new generator, known as **generalized self-shrinking generator** (GSSG), produces a family of sequences that has group structure. These sequences are also balanced and have quite good correlation.

2.4.1   Definition and Features

Let $\{a_i\}$, $i = 0, 1, 2, \ldots$, be a PN-sequence produced by an LFSR of L stages. Now, consider the binary vector

$$G = [g_0, g_1, \ldots, g_{L-1}] \in \mathbb{F}_2^L$$

and the sequence $\{v_i\}$, $i = 0, 1, 2, \ldots$, sometimes denoted by $v(G)$, such that

$$v_i = g_0 a_i + g_1 a_{i-1} + \cdots + g_{L-1} a_{i-L+1}.$$

Consider the following decimation rule:

$$\begin{cases} \text{If } a_i = 1, \text{ then } s_j = v_i. \\ \text{If } a_i = 0, \text{ then } v_i \text{ is discarded.} \end{cases}$$

This means that the PN-sequence $\{a_i\}$ decimates the sequence $\{v_i\}$, for each value of G.

We denote the sequence $\{s_j\}$, $j = 0, 1, 2, \ldots$, by $s(v)$ or $s(G)$ and call it **generalized self-shrunken sequence** (GSS-sequence). The family of GSS-sequences $s(a) = \{s(G) \mid G \in \mathbb{F}_2^L\}$ is the family of self-shrunken sequences based on the PN-sequence $\{a_i\}$.

It is worth noticing that the family of sequences

$$\left\{ \{v_i\}_{i \geq 0}, \mid v_i = g_0 a_i + g_1 a_{i-1} + \cdots + g_{L-1} a_{i-L+1}, G \in \mathbb{F}_2^L, G \neq 0 \right\}$$

includes all the $2^L - 1$ shifts sequences of $\{a_i\}$. Then, the PN-sequence $\{a_i\}$ decimates shift versions of itself.

For simplicity, we refer G as the decimal representation of the vector G.

Example 2.10 Consider the PN-sequence

$$\{a_i\} = \{1\ 1\ 1\ 0\ 0\ 1\ 0 \ldots\}$$

generated by the primitive polynomial $p(x) = 1 + x + x^3$. Since $\{a_i\}$ has period equal to 7, then we get 7 generalized self-shrunken sequences based on $\{a_i\}$ plus the identically zero sequence (see Table 2.1). ∎

Table 2.1 GSS-sequences generated by $1 + x + x^3$

G			v(G)							s(G)				LC
0	0	0	0	0	0	0	0	0	0	0	0	0	0	0
0	0	1	1	0	1	1	1	0	0	1	0	1	0	2
0	1	0	0	1	1	1	0	0	1	0	1	1	0	3
0	1	1	1	1	0	0	1	0	1	1	1	0	0	3
1	0	0	1	1	1	0	0	1	0	1	1	1	1	1
1	0	1	0	1	0	1	1	1	0	0	1	0	1	2
1	1	0	1	0	0	1	0	1	1	1	0	0	1	3
1	1	1	0	0	1	0	1	1	1	0	0	1	1	3
			1	1	1	0	0	1	0					

The family $s(a)$ is an L-dimensional linear space on $\mathbb{F}_2$, so it is an Abelian group with neutral element $\{0\,0\,\ldots\,0\,0\}$ [46]. Then, $|s(a)| = 2^L$.

The following results evidence the relation between some values of G and the generated sequences.

Theorem 2.5 ([46, Theorem 1])

1. $s(G) = \{0\,0\,0\,\ldots\}$ *if and only if* $G = [0, 0, \ldots, 0]$.
2. $s(G) = \{1\,1\,1\,\ldots\}$ *if and only if* $G = [1, 0, \ldots, 0]$.
3. $s(G)$ *is balanced otherwise.*

Theorem 2.6 ([46, Theorem 5])

1. *There are two sequences from* $s(G)$ *with period equal to 2, which are* $\{1\,0\,1\,0\ldots\}$ *and* $\{0\,1\,0\,1\,\ldots\}$.
2. *There are two sequences from* $s(G)$ *with period equal to 1, which are* $\{0\,0\,0\,0\ldots\}$ *and* $\{1\,1\,1\,1\,\ldots\}$.

This means that the identically 0 sequence, the identically 1 sequence and the sequences that alternate 0 and 1 belong to every family of GSS-sequences. Furthermore, the sequences different from the identically 0 sequence and the identically 1 sequence are balanced.

Theorem 2.7 ([10, Theorem 5]) *The characteristic polynomial of the GSS-sequences generated by a PN-sequence is* $p_{LC}(x) = (1 + x)^{LC}$, *where LC is the linear complexity of the considered GSS-sequence.*

Example 2.11 Consider the family of GSS-sequences obtained from the LFSR with characteristic polynomial $p(x) = 1 + x + x^3$ in Example 2.10. There are four different sequences (the others are shifted versions of these four) and it is possible to check, via the Berlekamp–Massey algorithm, that their corresponding characteristic polynomials are given by

$$\{0\} : 0\,0\,0\,0 \rightarrow p_0(x) = 1$$

$$\{4\} : 1\,1\,1\,1 \rightarrow p_1(x) = (1 + x)$$

$$\{1, 5\} : 1\,0\,1\,0 \rightarrow p_2(x) = (1 + x)^2$$

$$\{2, 3, 6, 7\} : 0\,1\,1\,0 \rightarrow p_3(x) = (1 + x)^3. \qquad \blacksquare$$

The generalized self-shrinking generator has hardly been studied. For example, there are no works on the period nor the complexity of the sequences. Since any PN-sequence possesses 2^{L-1} ones in its first period [41], it seems evident that the period of the sequences $s(G)$ is a power of 2, that is, 2^t with $t \leq L-1$. Again, due to experimental observations, we can claim that the period of the sequences different from the sequences mentioned in Theorem 2.6 is always 2^{L-1}. Furthermore, the linear complexity of these sequences seems to satisfy:

$$2^{L-2} < LC \leq 2^{L-1} - (L - 2).$$

The upper bound can be obtained adapting the proof given by Blackburn for the self-shrunken sequence in [5] . None of the other bounds has been proven yet. We encourage the reader to think about it.

Given a primitive polynomial $p(x)$ of degree L, the 2^{L-1} GSS-sequences generated are divided into $L - 1$ different groups depending on their LC:

- 1 sequence of $LC = 0$, the identically 0 sequence.
- 1 sequence of $LC = 1$, the identically 1 sequence.
- 2 sequences of $LC = 2$, sequences $\{0\ 1\ 0\ 1\ \ldots\}$ and $\{1\ 0\ 1\ 0\ \ldots\}$.
- 2^{i+1} sequences of linear complexity LC_i, with $2^{L-2} < LC_i \leq 2^{L-1} - (L - 2)$, for $i = 1, 2, \ldots, L - 2$, and $L_1 < L_2 < \cdots < L_{L-2}$.

Example 2.12 In Table 2.2, we can find the 32 GSS-sequences generated by $p(x) = 1 + x^2 + x^3 + x^4 + x^5$. There are:

- 1 sequence with $LC = 0$, the identically 0 sequence.
- 1 sequence with $LC = 1$, the identically 1 sequence.
- 2 sequences with $LC = 2$, sequences $\{0\ 1\ 0\ 1\ \ldots\}$ and $\{1\ 0\ 1\ 0\ \ldots\}$.
- 4 sequences with $LC = 10$.
- 8 sequences with $LC = 12$.
- 16 sequences with $LC = 13$.

∎

2.4.2   Generalized Self-Shrunken Sequences and Difference Equations

In this section we present the GSS-sequences as solutions of linear difference equations.

According to Theorem 2.7 and other results seen in the previous section, we know that the characteristic polynomial of the GSS-sequence generated by a maximal-length LSFR is of the form:

$$p_t(x) = (1 + x)^t, \quad t \leq 2^{L-1} - (L - 2).$$

This implies a linear recurrence relationship of the form:

$$(E + 1)^t s_j = 0, \tag{2.5}$$

with E being the one-sided shift introduced in expression (2.3). Expression (2.5) represents a linear binary constant coefficient difference equation whose characteristic polynomial $p_t(x)$ has a unique root $\lambda = 1$ with multiplicity t. The solutions of

Table 2.2 GSS-sequences generated by $p(x) = 1 + x^2 + x^3 + x^4 + x^5$

G	s(G)																LC
0	0	0	0	0	0	0	0	0	0	0	0	0	0	0	0	0	0
1	0	1	0	0	1	1	1	1	1	0	1	0	0	0	0	1	12
2	1	0	0	1	1	1	1	0	0	0	0	1	1	0	0	1	13
3	**1**	**1**	**0**	**1**	**0**	**0**	**0**	**1**	**1**	**0**	**1**	**1**	**1**	**0**	**0**	**0**	13
4	0	0	1	1	1	1	0	1	0	1	1	0	1	0	0	0	10
5	**0**	**1**	**1**	**1**	**0**	**0**	**1**	**0**	**1**	**1**	**0**	**0**	**1**	**0**	**0**	**1**	12
6	1	0	1	0	0	0	1	1	0	1	1	1	0	0	0	1	13
7	1	1	1	0	1	1	0	0	1	1	0	1	0	0	0	0	13
8	0	1	1	1	1	0	1	1	0	0	0	1	0	0	1	0	13
9	0	0	1	1	0	1	0	0	1	0	1	1	0	0	1	1	13
10	1	1	1	0	0	1	0	1	0	0	0	0	1	0	1	1	12
11	1	0	1	0	1	0	1	0	1	0	1	0	1	0	1	0	2
12	0	1	0	0	0	1	1	0	0	1	1	1	1	0	1	0	13
13	0	0	0	0	1	0	0	1	1	1	0	1	1	0	1	1	13
14	1	1	0	1	1	0	0	0	0	1	1	0	0	0	1	1	12
15	1	0	0	1	0	1	1	1	1	1	0	0	0	0	1	0	10
16	1	1	1	1	1	1	1	1	1	1	1	1	1	1	1	1	1
17	1	0	1	1	0	0	0	0	0	1	0	1	1	1	1	0	12
18	0	1	1	0	0	0	0	1	1	1	1	0	0	1	1	0	13
19	0	0	1	0	1	1	1	0	0	1	0	0	0	1	1	1	13
20	1	1	0	0	0	0	1	0	1	0	0	1	0	1	1	1	10
21	1	0	0	0	1	1	0	1	0	0	1	1	0	1	1	0	12
22	0	1	0	1	1	1	0	0	1	0	0	0	1	1	1	0	13
23	0	0	0	1	0	0	1	1	0	0	1	0	1	1	1	1	13
24	1	0	0	0	0	1	0	0	1	1	1	0	1	1	0	1	13
25	1	1	0	0	1	0	1	1	0	1	0	0	1	1	0	0	13
26	0	0	0	1	1	0	1	0	1	1	1	1	0	1	0	0	12
27	0	1	0	1	0	1	0	1	0	1	0	1	0	1	0	1	2
28	1	0	1	1	1	0	0	1	1	0	0	0	0	1	0	1	13
29	1	1	1	1	0	1	1	0	0	0	1	0	0	1	0	0	13
30	0	0	1	0	0	1	1	1	1	0	0	1	1	1	0	0	12
31	0	1	1	0	1	0	0	0	0	0	1	1	1	1	0	1	10

this equation are binary sequences $\{s_j\}$ whose generic term is given by

$$s_j = \binom{j}{0}c_0 + \binom{j}{1}c_1 + \cdots + \binom{j}{t-1}c_{t-1},$$

with $c_j \in \mathbb{F}_2$ and $\binom{j}{i}$ as binomial coefficients modulo 2, for $i = 0, 1, \ldots, t-1$ [59]. In fact, each binomial coefficient defines a succession of binary values with constant period T_j. Table 2.3 depicts the first binomial coefficients with their corresponding

Table 2.3 Binomial
coefficients reduced modulo
2, binary sequences and
periods

Bino. coeff.	Binary sequences	T_j
$\binom{j}{0}$	11111111	$T_0 = 1$
$\binom{j}{1}$	01010101	$T_1 = 2$
$\binom{j}{2}$	00110011	$T_2 = 4$
$\binom{j}{3}$	00010001	$T_3 = 4$
$\binom{j}{4}$	00001111	$T_4 = 8$
$\binom{j}{5}$	00000101	$T_5 = 8$
$\binom{j}{6}$	00000011	$T_6 = 8$
$\binom{j}{7}$	00000001	$T_7 = 8$

binary sequences and periods. The 2^t possible choices of c_i, $i = 0, 1, \ldots, t - 1$, provide the different binary sequences $\{s_j\}$ that satisfy expression (2.5). Particular choices of c_i give rise to the generalized self-shrunken sequences generated by an LFSR of L stages (including the SS-sequence and the MSS-sequence). Interestingly, all the solutions of expression (2.5) are the bit-wise sum of the basic sequences coming from the binomial coefficients (see Table 2.3) and weighted by c_i, $i = 0, 1, \ldots, t - 1$.

2.4.3 Relationship with the Modified Self-Shrinking Generator

In this section, we see how the MSS-sequence generated by a primitive polynomial $q(x)$ of degree L can be obtained as one of the GSS-sequences generated by another primitive polynomial of the same degree.

Theorem 2.8 ([10, Theorems 1–2]) *The MSS-sequence obtained by self-decimating a PN-sequence with characteristic polynomial $q(x)$ of degree L, with L odd, can be computed as one of the GSS-sequences using another primitive polynomial $p(x)$ of degree L given by*

$$p(x) = \left(x + \alpha^3\right)\left(x + \alpha^6\right)\left(x + \alpha^{12}\right) \cdots \left(x + \alpha^{3 \cdot 2^{L-1}}\right),$$

where $\alpha \in \mathbb{F}_{2^L}$ is a root of $q(x)$.

Notice that the self-shrunken sequence is also a generalized self-shrunken sequence [107]. When the PN-sequence $\{v_i\}$ is shifted 2^{L-1} bits regarding the PN-sequence $\{a_i\}$, then the generated sequence is the self-shrunken sequence.

Example 2.13 Given the LSFR with characteristic polynomial $q(x) = 1 + x^2 + x^5$ and the initial state $\{1\ 1\ 1\ 1\}$, we can obtain the following MSS-sequence:

$$\{1\ 1\ 0\ 0\ 1\ 0\ 0\ 1\ 0\ 1\ 1\ 1\ 0\ 0\ 1\ 0 \ldots\}.$$

According to Theorem 2.8, this sequence can be also obtained using the GSSG with primitive polynomial

$$p(x) = \left(x + \alpha^3\right)\left(x + \alpha^6\right)\left(x + \alpha^{12}\right)\left(x + \alpha^{24}\right)\left(x + \alpha^{48}\right) = 1 + x^2 + x^3 + x^4 + x^5,$$

where $\alpha \in \mathbb{F}_{2^5}$ is a root of $q(x)$. In Table 2.2 we can find the 32 GSS-sequences generated by $p(x)$ using the different values of G. For $G = 5$ ($G = [1\,0\,1\,0\,0]$), the generated GSS-sequence is a shifted version of the MSS-sequence generated by $q(x)$.

The self-shrunken sequence generated by $p(x) = 1 + x^2 + x^3 + x^4 + x^5$ is also a GSS-sequence. For instance, consider the initial state $\{1\,1\,1\,1\,1\}$, we generate the following SS-sequence:

$$\{1\,1\,0\,1\,0\,0\,0\,1\,1\,0\,1\,1\,1\,0\,0\,0\,\ldots\},$$

which is exactly the GSS-sequence corresponding to $G = 3$ ($G = [0\,0\,0\,1\,1]$) (see Table 2.2). ∎

Now, in order to know which GSS-sequence is the MSS-sequence, we need to recall the definition of Zech logarithm. Zech logarithms are named after Julius Zech who published in 1849 a table of this type logarithms (which he called *addition logarithms*) for doing arithmetic in $\mathbb{Z}_p$. These logarithms are also called as Jacobi logarithms after C.G.J. Jacobi who used them for number theoretic investigations [48].

Definition 2.1 Let $\mathbb{F}_q$ be the Galois field of q elements and $\alpha \in \mathbb{F}_q$ a primitive element. The **Zech logarithm** with basis α is the application $\mathscr{Z}_\alpha : \mathbb{Z}_q \to \mathbb{Z}_q^* \cup \{\infty\}$, such that each element $t \in \mathbb{Z}_q$ corresponds to $\mathscr{Z}_\alpha(t)$, attaining $1 + \alpha^t = \alpha^{\mathscr{Z}_\alpha(t)}$.

Now we are ready to compute the value of G that produces a MSS-sequence as a GSS-sequence.

Theorem 2.9 ([10, Theorem 3]) *The MSS-sequence generated from a PN-sequence with primitive characteristic polynomial $q(x)$ is also a GSS-sequence obtained from a PN-sequence generated by a primitive polynomial $p(x)$ (see Theorem 2.8) that decimates a shifted version of itself with shift $(D - 2)3^{-1} \bmod (2^L - 1)$, where $D = \mathscr{Z}_\alpha(1)$, $\alpha \in \mathbb{F}_{2^L}$ is a root of $p(x)$ and L is the degree of $p(x)$ and $q(x)$.*

Assume $\{a_i\}$ is a PN-sequence generated by a primitive polynomial and assume $\{b_i\} = \{a_{i+(D-2)3^{-1}}\}$ is the shifted version of $\{a_i\}$ that is decimated by $\{a_i\}$ in order to obtain the MSS-sequence (see Theorem 2.9). According to the definition of GSSG, to find the value of $G = [g_0, g_1, \ldots, g_{L-1}]$ that generates $\{b_i\}$ from $\{a_i\}$, we have to know L bits of $\{a_i\}$ and $2L - 1$ bits of $\{b_i\}$ to solve the following system:

$$
\begin{cases}
a_0 = b_0 g_0 + b_{2^L-2} g_1 + b_{2^L-3} g_2 + \cdots + b_{2^L-L} g_{L-1} \\
a_1 = b_1 g_0 + b_0 g_1 + b_{2^L-2} g_2 + \cdots + b_{2^L-(L-1)} g_{L-1} \\
\quad \vdots \\
a_{L-1} = b_{L-1} g_0 + b_{L-2} g_1 + b_{L-3} g_2 + \cdots + b_0 g_{L-1}.
\end{cases} \tag{2.6}
$$

The exact necessary bits of each sequence are $\{a_i\}_{i=0}^{L-1}$ and $\{b_i\}_{i=0}^{L-1} \cup \{b_i\}_{i=2^L-L}^{2^L-2}$, respectively.

Let us see a clarifying example.

Example 2.14 Consider the MSS-sequence generated in Example 2.13:

$$\{s_j\} = \{1\,1\,0\,0\,1\,0\,0\,1\,\underline{0}\,1\,1\,1\,0\,0\,1\,0\,\ldots\}. \tag{2.7}$$

According to Theorem 2.8, $\{s_j\}$ can be generated as a GSS-sequence using the primitive polynomial $p(x) = 1 + x^2 + x^3 + x^4 + x^5$. Given the PN-sequence $\{a_i\}$ generated by $p(x)$, we consider the PN-sequence $\{b_i\} = \{a_{i+k}\}$ which is a shifted version of $\{a_i\}$, with shift $k = (\mathscr{L}_\alpha(1) - 2) \cdot 3^{-1} \bmod 31$, α root of $q(x)$. This means that $\{b_i\} = \{a_{i+26}\}$.

Taking the initial state $\{1\,1\,1\,1\,1\}$, we can generate the PN-sequence

$$\{a_i\} = \{1\,1\,1\,1\,1\,0\,1\,1\,1\,0\,0\,0\,1\,0\,1\,0\,1\,1\,0\,1\,0\,0\,0\,0\,1\,1\,0\,0\,1\,0\,0\ldots\}$$

that decimates a shifted version of itself,

$$\{b_i\} = \{\mathbf{0}\,\mathbf{1}\,\mathbf{1}\,\mathbf{1}\,0\,0\,0\,1\,0\,1\,0\,1\,1\,0\,1\,0\,0\,0\,0\,1\,1\,0\,0\,1\,0\,0\,\mathbf{1}\,\mathbf{1}\,\mathbf{1}\,\mathbf{1}\,\mathbf{1}\},$$

with shift equal to 26. Thus, we obtain the output sequence

$$\{0\,1\,1\,1\,0\,0\,1\,0\,1\,1\,0\,0\,1\,0\,0\,1\ldots\},$$

which is a shifted version of the MSS-sequence $\{s_j\}$ starting at the underlined position (see expression (2.7)).

Now, in order to obtain the value of $G = [g_0, g_1, g_2, g_3, g_4]$, we have to solve the system given in (2.6). In this case we have

$$\{a_i\}_{i=0}^{4} = \{0\,1\,1\,1\,0\}, \quad \{b_i\}_{i=0}^{4} = \{1\,1\,1\,1\,1\} \quad \text{and} \quad \{b_i\}_{i=27}^{30} = \{0\,1\,0\,0\}.$$

Table 2.4 GSS-sequences obtained with $G = 5$

Initial state					5th GSS-sequence															
1	1	1	1	1	0	1	1	1	0	0	1	0	1	1	0	0	1	0	0	1
1	0	0	0	0	1	0	0	1	0	1	1	1	0	0	1	0	1	1	0	0
1	0	1	0	1	1	1	0	0	1	0	0	1	0	1	1	1	0	0	1	0

Therefore, system (2.6) has the following form:

$$\begin{cases} 0 = g_0 + g_3 \\ 1 = g_0 + g_1 + g_4 \\ 1 = g_0 + g_1 + g_2 \\ 1 = g_0 + g_1 + g_2 + g_3 \\ 0 = g_0 + g_1 + g_2 + g_3 + g_4, \end{cases}$$

whose solution is $G = [0\ 0\ 1\ 0\ 1]$ ($G = 5$). Then, the GSSG with primitive polynomial $p(x) = 1 + x^2 + x^3 + x^4 + x^5$ produces the MSS-sequence $\{s_j\}$ for $G = 5$ ($G = [0\ 0\ 1\ 0\ 1]$) for any given initial state. For example, in Table 2.4, we can see that the GSS-sequence produced with $G = 5$ using three different initial states provides shifted versions of the same sequence, the MSS-sequence $\{1\ 1\ 0\ 0\ 1\ 0\ 0\ 1\ 0\ 1\ 1\ 1\ 0\ 0\ 1\ 0\}$. ∎

Chapter 3
Modelling Through Linear Cellular Automata

The irregular decimation was introduced to break the linearity of the PN-sequences. However, in this chapter we will see that there exist linear structures that describe the behaviour of the shrinking generators, designed as non-linear. The inherent linearity of these structures can be used to cryptanalyse such generators as described in Chap. 4.

3.1 The Concept of Cellular Automaton

Cellular automata (CAs) are particular forms of finite state machines defined as uniform arrays of identical cells in an n-dimensional state. A cellular automaton (CA) evolves in discrete time steps, within the content of one cell being affected by the contents of cells in its neighbourhood on the previous time step. That is, the value of the ith cell at time $t + 1$, denoted by x_i^{t+i}, depends on the contents of the k neighbour cells at time t.

One-dimensional CAs with $k = 3$ and with contents in the binary field are called elementary CAs. There are 2^3 possible configurations for each cell and its two immediate neighbours. The rule defining the cellular automaton must specify the resulting state for each of these possibilities so there are 2^{2^3} possible rules for elementary CA evolution. These rules can be considered as Boolean functions.

Stephen Wolfram proposed a scheme, known as the Wolfram code, to assign each rule a number from 0 to 255 [103]. Each possible current configuration of three neighbour cells is written in the order, 111, 110, ... , 001, 000, and the resulting state for each configuration is written in the same order and interpreted as the binary

S. Díaz Cardell, A. Fúster-Sabater, *Cryptography with Shrinking Generators*,
SpringerBriefs in Mathematics, https://doi.org/10.1007/978-3-030-12850-0_3

representation of an integer. For example, one can find the four rules we will use in this work below:

Rule 150: $x_i^{t+1} = x_{i-1}^t + x_i^t + x_{i+1}^t$

111	110	101	100	011	010	001	000
1	0	0	1	0	1	1	0

Rule 90: $x_i^{t+1} = x_{i-1}^t + x_{i+1}^t$

111	110	101	100	011	010	001	000
0	1	0	1	1	0	1	0

Rule 102: $x_i^{t+1} = x_i^t + x_{i+1}^t$

111	110	101	100	011	010	001	000
0	1	1	0	0	1	1	0

Rule 60: $x_i^{t+1} = x_{i-1}^t + x_i^t$

111	110	101	100	011	010	001	000
0	0	1	1	1	1	0	0

Notice that 10010110, 01011010, 01100110 and 00111100 are the binary representations of 150, 90, 102 and 60, respectively.

Many of the rules seem to generate patterns with evident structures. For example, Fig. 3.1 shows the AC-images generated by these four rules after applying 15 iterations to the one-dimensional CA. One can notice the symmetry between rules 60 and 102 and that both rules generate a fractal structure. Rules 150 and 90 produce symmetric structures and are both additive rules. Every additive rule is able to emulate itself and produce nested patters [103].

Observe, for example, Rules 30 and 94, which are non-linear:

Rule 30: $x_i^{t+1} = x_{i-1}^t + x_i^t + x_{i+1}^t + x_i^t x_{i+1}^t$

111	110	101	100	011	010	001	000
0	0	0	1	1	1	1	0

Rule 94: $x_i^{t+1} = x_{i-1}^t + x_i^t + x_{i+1}^t + x_{i-1}^t x_i^t + x_i^t x_{i+1}^t + x_{i-1}^t x_i^t x_{i+1}^t$

111	110	101	100	011	010	001	000
0	1	0	1	1	1	1	0

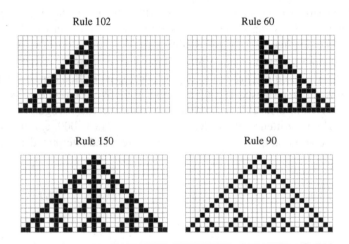

Rule 102　　　　　　　　　　　　　　Rule 60

Rule 150　　　　　　　　　　　　　　Rule 90

Fig. 3.1 AC-images generated with Rules 102, 60, 150 and 90

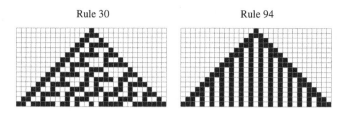

Fig. 3.2 AC-images generated by Rules 30 and 94

The AC-image generated by Rule 30 shows no recognizable pattern (see Fig. 3.2). On the other hand, Rule 94 is an example of simple CA whose evolution corresponds to computations that can be easily described in traditional mathematical terms. Patterns can show both for linear rules (e.g., Rules 60 and 102) and for non-linear rules (e.g., Rule 94).

When the rules involved in the CA use only XOR operations, the CA is said to be **linear**. Notice that Rules 60, 102, 150 and 90 use only XOR operations. This means that we will only consider linear CAs in this chapter.

Due to their capability to exhibit complex behaviours, CAs have applications in many different areas, for example, in modelling physical systems [58, 100] and non-linear chemical systems [102], studying problems of number theory [86, 102] or as pseudorandom number generators [97].

Furthermore, due to the speed and randomness in their sequences, CAs are a very good basis for stream ciphers. What is more, their hardware implementation is simple and their regular structure makes possible to find an efficient software implementation. The first cryptographic application of CAs was published in [101]. In this work, Wolfram used Rule 30 for building a stream cipher that was afterwards broken by Meier and Stafflebach [66]. Besides, other authors have proposed stream ciphers based on CAs along the years [20, 51, 73].

Next, we classify the elementary CAs.

Definition 3.1 An elementary CA is said to be:

- **Uniform** or **regular** if every cell is computed using the same rule.
- **Hybrid** if different rules are considered when computing the contents of the cells.
- **Null** if cells with null content are adjacent to the extreme cells when it is needed.
- **Periodic** or **cyclic** if extreme cells are adjacent.

In Table 3.1a we can find an example of a regular, cyclic 102-CA of length 3. Furthermore, since Rule 102 only operates the contents of a cell and its right neighbour cell, we consider cyclic boundary only on the right of the CA in order to compute the last vertical sequence. Given the initial state {1 1 0}, the CA generates as many new states of length 3 as we want in the following way:

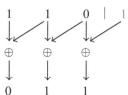

Table 3.1 Examples of elementary CAs

(a) Regular cyclic 102-CA				(b) Regular cyclic 60-CA				(c) Hybrid null 150/90-CA				
102	102	102			60	60	60		90	150	150	90
1	1	0	1	1	0	1	1	0	1	0	1	0	0
0	1	1	0	0	1	1	0	0	0	0	1	1	0
1	0	1	1	1	1	0	1	0	0	1	0	1	0
1	1	0	1	1	0	1	1	0	1	1	0	0	0
0	1	1	0	0	1	1	0	0	1	0	1	0	0
1	0	1	1	1	1	0	1	0	0	0	1	1	0
1	1	0	1	1	0	1	1	0	0	1	0	1	0
0	1	1	0	0	1	1	0	0	1	1	0	0	0
⋮	⋮	⋮			⋮	⋮	⋮		⋮	⋮	⋮	⋮	

However, at some point, these states start to recur; thus, the CA generates 3 (vertical) output sequences with period 3.

In Table 3.1b, we find a regular, cyclic 60-CA of length 3. Since Rule 60 only operates the contents of a cell and its left neighbour cell, in this case we consider cyclic boundary only on the left of the CA. Note that the (vertical) output sequences generated by this 60-CA are the same (vertical) sequences generated in the 102-CA in Table 3.1a (they appear in inverse order).

Finally, in Table 3.1c, we can find one example of hybrid null 150/90-CA of length 4. In this case, we have to consider null boundary in both sides of the CA. Besides, the CA generates four (vertical) output sequences with period 7.

In general, the (vertical) sequences generated by a 102-CA (60-CA) have different periods. In addition, due to the symmetry between rules 102 and 60, the sequences generated by a 102-CA of length L can be also generated by a 60-CA of length L.

3.2 Modelling a PN-Sequence

In this section we will see how to obtain PN-sequences by means of elementary linear CAs. We recall that a PN-sequence is a sequence generated by an LFSR whose characteristic polynomial is primitive.

3.2.1 Cattell–Muzio Algorithm

In [13], Cattell and Muzio presented a method for computing a 90/150-CA that generates the same sequences as those produced by a given irreducible characteristic polynomial. This approach is based on a correspondence between the characteristic polynomial calculations and GCD computations. In fact, they proved that each irreducible polynomial has exactly two CA realizations.

First of all we need to recall the definition of trace of a polynomial.

Definition 3.2 ([13]) The trace of a polynomial $q(x)$ with respect to an irreducible polynomial $p(x)$ of degree L is given by

$$\mathrm{Tr}(q(x)) = \left[q(x) + q(x)^2 + q(x)^4 + \cdots + q(x)^{2^{L-1}} \right] \bmod p(x).$$

It is worth noticing that the trace of a polynomial is always zero or one.

Example 3.1 Consider the polynomial $p(x) = x^2$ and the primitive polynomial $p(x) = 1 + x^2 + x^5$. First, we compute the powers of $q(x)$:

$$q(x)^2 = x^4$$

$$q(x)^4 = x^8 \bmod p(x) = 1 + x^2 + x^3$$

$$q(x)^8 = \left(1 + x^4 + x^6\right) \bmod p(x) = 1 + x + x^3 + x^4$$

$$q(x)^{16} = \left(1 + x^2 + x^6 + x^8\right) \bmod p(x) = x.$$

Summing these polynomials, we find the trace of $q(x)$:

$$\mathrm{Tr}(q(x)) = q(x) + q(x)^2 + q(x)^4 + q(x)^8 + q(x)^{16} = 0.$$

∎

The method given in Algorithm 1 shows the necessary process to compute a CA for a given irreducible characteristic polynomial $p(x)$ of degree L. This algorithm is very easy to code in languages such as Maple, Python, etc. As a consequence of Algorithm 1, we can introduce the following result.

Theorem 3.1 ([28]) *For a PN-sequence generated by a primitive polynomial of degree L, there exists an hybrid, null 150/90-CA of length L that generates such PN-sequence.*

Example 3.2 Consider the primitive polynomial $p(x) = 1 + x^2 + x^5$. Since primitive implies irreducible, we can apply Algorithm 1.

Algorithm 1 Cattell–Muzio algorithm

Input: An irreducible polynomial $p(x)$
01:   Compute $f(x) = (x^2 + x) p'(x)$
02:   Compute $g(x) = (1/f(x))^2$
03:   if L is even
04:       Find $\theta(x)$ with trace 1
05:       Compute $\beta(x) = \sum_{i=1}^{L-1} \left(\sum_{j=0}^{i-1} g^{2^j} \right) \theta^{2^i}$
06:   else
07:       Compute $\beta(x) = \sum_{i=1}^{(L-1)/2} g^{2^{2i-1}}$
08:   endif
09:   $q(x) = \beta(x) f(x)$
10:   Compute $\gcd(p(x), q(x))$, saving the quotients
11:   Construct the CA from the constant terms of the quotients
Output:
        A binary string of length L codifying a CA corresponding to the PN-sequence generated by $p(x)$

We compute the derivative of $p(x)$ modulo 2:

$$p'(x) \bmod 2 = \left(2x + 5x^4 \right) \bmod 2 = x^4.$$

Now we compute $f(x)$ modulo $p(x)$:

$$f(x) = \left(x + x^2 \right) p'(x) = \left(x^5 + x^6 \right) \bmod p(x) = 1 + x + x^2 + x^3.$$

Next, we use the extended Euclidean GCD algorithm to compute the inverse of $f(x)$:

$$1/f(x) = 1 + x^2 + x^3.$$

We compute $g(x)$

$$g(x) = (1/f(x))^2 = \left(1 + x^4 + x^6 \right) \bmod p(x) = 1 + x + x^3 + x^4$$

and the powers of $g(x)$:

$$g^2(x) = \left(1 + x^2 + x^6 + x^8 \right) \bmod p(x) = x$$

$$g^4(x) = x^2$$

$$g^8(x) = x^4$$

$$g^{16}(x) = x^8 \bmod p(x) = x^3 + x^2 + 1.$$

Summing $g(x)$ and its powers we get that the trace of $g(x)$ with respect to $p(x)$ is zero.

Since $L = 5$ odd, we compute $\beta(x)$ as follows:

$$\beta(x) = \sum_{i=1}^{2} g(x)^{2^{2i-1}} = g(x)^2 + g(x)^8 = x + x^4.$$

Finally,

$$q(x) = \left(x + x^2\right) p'(x)\beta(x)$$
$$= \left(x + x^2\right) x^4 \left(x + x^4\right)$$
$$= 1 + x^2 + x^4.$$

Now, we apply the Euclid's algorithm to search $\gcd(p(x), q(x))$:

$$1 + x^2 + x^5 = \left(1 + x^2 + x^4\right) x + \left(1 + x + x^2 + x^3\right)$$
$$1 + x^2 + x^4 = (1 + x)\left(1 + x + x^2 + x^3\right) + x^2$$
$$1 + x + x^2 + x^3 = (1 + x)x^2 + (1 + x)$$
$$x^2 = (1 + x)(1 + x) + 1$$
$$1 + x = (1 + x)1 + 0.$$

This process returns the quotients

$$[x, 1 + x, 1 + x, 1 + x, 1 + x]$$

and so the CA is constructed from the constant terms

$$[0, 1, 1, 1, 1]. \tag{3.1}$$

Now, we substitute 0 and 1 by 90 and 150, respectively. Thus the CA given by [90, 150, 150, 150, 150] generates the PN-sequences produced by $p(x)$.

Now, we consider the mirror image of (3.1)

$$[1, 1, 1, 1, 0]$$

that represents the CA

$$[150 \ 150 \ 150 \ 150 \ 90]$$

that also generates the PN-sequences produced by $p(x)$.

Table 3.2 Null 105/90-CA
that generates the
PN-sequence produced by
$p(x) = 1 + x^2 + x^5$

150	150	150	150	90
1	0	1	0	1
1	0	1	0	0
1	0	1	1	0
1	0	0	0	1
1	1	0	1	0
0	0	0	1	1
0	0	1	0	1
0	1	1	0	0
1	0	0	1	0
1	1	1	1	1
0	1	1	1	1
1	0	1	1	1
1	0	0	1	1
1	1	1	0	1
0	1	0	0	0
1	1	1	0	0
0	1	0	1	0
1	1	0	1	1
0	0	0	0	1
0	0	0	1	0
0	0	1	1	1
0	1	0	1	1
1	1	0	0	1
0	0	1	1	0
0	1	0	0	1
1	1	1	1	0
0	1	1	0	1
1	0	0	0	0
1	1	0	0	0
0	0	1	0	0
0	1	1	1	0

For instance, consider the PN-sequence

$$\{1\ 1\ 1\ 1\ 1\ 0\ 0\ 0\ 1\ 1\ 0\ 1\ 1\ 1\ 0\ 1\ 0\ 1\ 0\ 0\ 0\ 0\ 1\ 0\ 0\ 1\ 0\ 1\ 1\ 0\ 0 \ldots\}$$

generated by $p(x)$ with initial state $\{1\ 1\ 1\ 1\ 1\}$, this sequence can be generated as well by the 150/90-CA given in Table 3.2.   ■

The algorithm is sufficiently fast for practical applications and the number of operations does not depend on the input polynomial, only on its degree.

3.2.2 Other CAs that Generate PN-Sequences

In this section, we show that for every PN-sequence there also exists a 102-CA that generates it.

We start the section with an important result about PN-sequences that will be needed afterwards.

Theorem 3.2 ([9, Theorem 3.6]) *For a PN-sequence $\{a_i\}$ generated by a primitive polynomial $p(x)$ of degree L, there exists a unique number $D \in \{2, 3, \ldots, 2^L - 2\}$ such that $a_i + a_{i+1} = a_{i+D}$. This number is $D = \mathcal{Z}_\alpha(1)$, with $\alpha \in \mathbb{F}_{2^L}$ a root of $p(x)$.*

It is worth recalling that $\mathcal{Z}_\alpha(1)$ is the Zech logarithm of 1 in basis α (see Definition 2.1).

According to Theorem 3.2 and the general form of a 102-CA (see Table 3.3), if the PN-sequence $\{a_i\}$ appears in the 0th column of a 102-CA, the other sequences are shifted versions of such PN-sequence. Furthermore, the sequence in the tth column it is a shifted version of $\{a_i\}$, that is, $\{a_{i+d}\}$, with $d = t \cdot D \mod (2^L - 1)$. Eventually, the PN-sequence $\{a_i\}$ itself will appear again; thus, the 102-CA has finite length. The general form of the columns of a 102-CA can be found in Sect. 4.6.1 (Method 2).

Next result, whose proof is left as an exercise, claims that given a primitive polynomial there always exists a 102-CA that generates the PN-sequence produced by such polynomial.

Theorem 3.3 *There exists a regular, cyclic 102-CA of length $\frac{2^L - 1}{\gcd(D, 2^L - 1)}$, with D as in Theorem 3.2 that generates the same PN-sequence as that produced by a primitive polynomial $p(x)$ of degree L.*

As an example, consider the PN-sequence generated by $p(x) = 1 + x^2 + x^5$ given in Example 3.2. There exists a regular, cyclic 102-CA of length 31 that generates such PN-sequence (see Table 3.4). What is more, all the sequences are shifted versions of the same PN-sequence. Since the characteristic polynomial of the PN-sequence is $p(x) = 1 + x^2 + x^5$, it is easy to check that $D = 18$. This means that the shift from one sequence to the following is 18. For example, the sequence in the first column is a shifted version of the PN-sequence in the 0th column, but starting

Table 3.3 General form of a 102-CA

102	102	102	102	102	...	102	...
a_0	$a_0 + a_1$	$a_0 + a_2$	$a_0 + a_1 + a_2 + a_3$	$a_0 + a_4$	...	$a_0 + a_8$	...
a_1	$a_1 + a_2$	$a_1 + a_3$	$a_1 + a_2 + a_3 + a_4$	$a_1 + a_5$	...	$a_1 + a_9$	...
a_2	$a_2 + a_3$	$a_2 + a_4$	$a_2 + a_3 + a_4 + a_5$	$a_2 + a_6$	...	$a_2 + a_{10}$	...
a_3	$a_3 + a_4$	$a_3 + a_5$	$a_3 + a_4 + a_5 + a_6$	$a_3 + a_7$	...	$a_3 + a_{11}$	...
$\vdots$	$\vdots$	$\vdots$	$\vdots$	$\vdots$		$\vdots$	

Table 3.4 102-CA that generates the PN-sequence produced by $p(x) = 1 + x^2 + x^5$

in position 18 (see circled bits in Table 3.4). The sequence in the second column is a shifted version of the PN-sequence in the 0th column but starting in position $2 \cdot 18 \bmod 31$, that is, in position 5 (see squared bits in Table 3.4) and so on.

Note that when $\gcd(D, 2^L - 1) = 1$, the length of the 102-CAs mentioned in Theorem 3.3 is $2^L - 1$. The length of the 150/90-CAs proposed in Sect. 3.2.1 is much smaller. However, if we know $p(x)$ and the PN-sequence $\{a_i\}$, we can compute D as in Theorem 3.2 and we can complete the 102-CA with the corresponding shifted versions of $\{a_i\}$. In addition, since the 102-CA proposed in Theorem 3.3 is regular, every cell follows the same rule and the form of the CA is immediately obtained. On the other hand, in order to find the form of the 150/90-CA in Sect. 3.2.1, we need to apply the Cattell–Muzio Algorithm [13].

3.3 Modelling the Shrinking Generator

In this section, we present two different families of linear CAs that generate the shrunken sequence produced by two maximum-length LFSRs. From now on, we denote by $p_1(x)$ and $p_2(x)$ of degrees L_1 and L_2, the primitive characteristic polynomials of such registers, respectively.

3.3.1 The Fúster–Caballero Algorithm

In [28] the authors proposed an algorithm that provides a 150/90-CA that generates the shrunken sequence produced by two maximum-length LFSRs. This approach is based on the Cattell–Muzio Algorithm [13] seen in Sect. 3.2.1 and a CA-concatenation technique.

Algorithm 2 provides two hybrid, null 150/90-CAs that produce the shrunken sequence generated by $p_1(x)$ and $p_2(x)$. Actually, the algorithm is based on the concatenation of the CA produced applying the Cattell–Muzio algorithm for $p_2(x)$ [13].

Notice that $p_1(x)$ only contributes the number of concatenations according to its degree. This polynomial is no further implicated in the algorithm, this means that with $p_2(x)$ fixed, for different values of $p_1(x)$ with degree L_1 the algorithm would return the same result.

Example 3.3 Consider the primitive polynomial $p_2(x) = 1 + x + x^2 + x^4 + x^5$ and a primitive polynomial $p_1(x)$ of degree 3.

First, we compute $N = 1 + 2 + 4 = 7$ and the polynomial

$$p(x) = \left(x + \alpha^7\right)\left(x + \alpha^{14}\right)\left(x + \alpha^{28}\right)\left(x + \alpha^{56}\right)\left(x + \alpha^{112}\right) = 1 + x^2 + x^5,$$

where α is a primitive element of $\mathbb{F}_{2^5}$, root of $p_2(x)$.

Algorithm 2 Fúster–Caballero algorithm

Input: L_1 and $p_2(x)$

01:   Compute $N = 2^0 + 2^1 + 2^2 + \cdots + 2^{L_1-1}$

02:   Compute $p(x) = \left(x + \alpha^N\right)\left(x + \alpha^{2N}\right) \cdots \left(x + \alpha^{2^{L_2-1}N}\right)$, with α root of $p_2(x)$

03:   Compute two linear 90/150 CA, denoted by s_i, $i = 1, 2$, for $p(x)$ using the Cattell-Muzio algorithm

04:   for $j = 1 : L_1 - 1$

05:       Complement the last bit of s_i and denote the resultant string as t_i

06:       Compute de mirror image of t_i, denoted by t_i^* and concatenate both strings: $s_i = t_i t_i^*$

07:   endfor

Output:

     Two binary strings of length $L_2 \cdot 2^{L_1-1}$ codifying two CAs corresponding to the shrinking generator

---

Now, applying the Cattell–Muzio algorithm to $p(x)$, we obtain two strings that represent two 150/90-CAs (see Example 3.2):

$$[01111] \rightarrow [90\ 150\ 150\ 150\ 150]$$

$$[11110] \rightarrow [150\ 150\ 150\ 150\ 90].$$

We choose, for example, the first CA and we perform the concatenation process $L_1 - 1 = 2$ times:

$$[01111]$$
$$[0111001110]$$
$$[01110011111111001110].$$

For the second CA, we proceed in the same manner:

$$[11110]$$
$$[1111111111]$$
$$[11111111100111111111].$$

Now, we substitute 0 and 1 by 90 and 150, respectively, and we obtain two CAs

$$[90\ 150\ 150\ 150\ 90\ 90\ 150\ 150\ 150\ 150\ 150\ 150\ 150\ 150\ 90\ 90\ 150\ 150\ 150\ 90]$$

$$[150\ 150\ 150\ 150\ 150\ 150\ 150\ 150\ 150\ 90\ 90\ 150\ 150\ 150\ 150\ 150\ 150\ 150\ 150\ 150],$$

both of them capable of generating the shrunken sequence generated by $p_2(x)$ and $p_1(x)$. ∎

3.3.2 Other CAs that Generate the Shrunken Sequence

In [9], the authors proposed a family of 102-CAs (60-CAs) that also generate the shrunken sequence.

Again, consider two primitive polynomials $p_1(x)$ and $p_2(x)$ of length L_1 and L_2, respectively. We can introduce the following result.

Theorem 3.4 ([9, Theorem 3.10]) *The shrunken sequence generated by $p_1(x)$ and $p_2(x)$ can be generated by a regular, cyclic 102-CA of length $\frac{T}{\gcd(2^{L_2}-1, D)}$, where $D = \mathcal{L}_\alpha(1)$, with $\alpha \in \mathbb{F}_{2^{L_2}}$ root of $p(x)$ (see Theorem 2.1) and $T = 2^{L_1-1}(2^{L_2} - 1)$ is the period of the shrunken sequence.*

Apart from the shrunken sequence, other $2^{L_1-1} - 1$ sequences, *the companion sequences*, with the same period and characteristic polynomial as those of the shrunken sequence are generated by the 102-CA [9]. Furthermore, shifted versions of these sequences, including the shrunken sequence, appear along the 102-CA.

Notice that the sequences in columns $t \cdot 2^{L_1-1}$, with $t = 1, 2, \ldots, L/(2^{L_1-1}-1)$, are shifted versions of the shrunken sequence, with shift equal to $t \cdot D \cdot 2^{L_1-1}$, for $t = 1, 2, \ldots, L/(2^{L_1-1} - 1)$ [9, 11]. Moreover, the companion sequence in the column $t \cdot 2^{L_1-1}+m$, for $m = 1, 2, \ldots, 2^{L_1-1} - 1$ and $t = 0, 1, \ldots, L/(2^{L_1-1})-1$, is a shifted version of the companion sequence in the mth column starting in position $t \cdot D \cdot 2^{L_1-1}$[11].

Example 3.4 Consider the shrunken sequence generated by $p_1(x) = 1 + x + x^2$ and $p_2(x) = 1 + x + x^3$, in Example 2.2:

$$\{1\,0\,1\,1\,1\,0\,0\,0\,1\,1\,0\,1\,0\,1\ldots\}.$$

This sequence has characteristic polynomial $p(x)^2 = \left(1 + x^2 + x^3\right)^2$ and period $T = 14$. In Table 3.5 there is an example of a regular, cyclic 102-CA of length 14 that generates this sequence in the 0th column. This CA generates 2 different sequences, the shrunken sequence and one companion sequence, both with the same period and characteristic polynomial. Shifted versions of these two sequences appear 7 times along the 102-CA: the shrunken sequence appears in columns 0, 2, 4, 6, 8, 10 and 12, and the companion sequence appears in columns, 1, 3, 5, 7, 9, 11 and 13.

Now, we can compute $2^{L_1-1} = 2$ and $D = \mathcal{L}_\alpha(1) = 5$, with $\alpha \in \mathbb{F}_{2^3}$ root of $p(x)$. We consider, for example, the 2nd column of the 102-CA. In this case $t = 1$, therefore this sequence is a shifted version of the shrunken sequence with shift equal to $2^{L_1-1} \cdot t \cdot D \bmod 14 = 10$ (see the circled bit of the shrunken sequence in Table 3.5). Consider now, for example, the 9th column of the 102-CA. Since now $t = 4$, the considered sequence is a shifted version of the companion sequence with shift equal to $2^{L_1-1} \cdot t \cdot D \bmod 14 = 12$ (see the squared bit of the shrunken sequence in Table 3.5). ∎

In Sect. 2.1.2, we saw that the shrunken sequence is composed of interleaving shifted versions of a PN-sequence generated by the primitive polynomial $p(x)$.

Table 3.5 CA that generates the shrunken sequence in Example 2.2

102	102	102	102	102	102	102	102	102	102	102	102	102	102
1	1	(0)	1	0	0	1	0	0	[1]	1	0	1	1
0	1	1	1	0	1	1	0	1	0	1	1	0	0
1	0	0	1	1	0	1	1	1	1	0	1	0	0
1	0	1	0	1	1	0	0	0	1	1	1	0	1
1	1	1	1	0	1	0	0	1	0	0	1	1	0
0	0	0	1	1	1	0	1	1	0	1	0	1	1
0	0	1	0	0	1	1	0	1	1	1	1	0	1
0	1	1	0	1	0	1	1	0	0	0	1	1	1
[1]	0	1	1	1	1	0	1	0	0	1	0	0	1
1	1	0	0	0	1	1	1	0	1	1	0	1	0
(0)	1	0	0	1	0	0	1	1	0	1	1	1	1
1	1	0	1	1	0	1	0	1	1	0	0	0	1
0	1	1	0	1	1	1	1	0	1	0	0	1	0
1	0	1	1	0	0	0	1	1	1	0	1	1	0

As a consequence of the formation rule of the 102-CA and the fact that summing elements of a PN-sequence generates another PN-sequence [41], it is possible to check that the companion sequences are also composed of interleaving shifted versions of the same PN-sequence. We leave this claim as an exercise for the reader.

Let us denote the interleaved PN-sequences of the shrunken sequence by $\left\{v_{d_0^0+i}\right\}, \left\{v_{d_1^0+i}\right\}, \left\{v_{d_2^0+i}\right\}, \ldots, \left\{v_{d_{2^{L_1-1}-1}^0+i}\right\}, i = 0, 1, \ldots,$ where $d_0^0 = 0$. Remember that the positions d_k^0 depend on the location of the 1s in the PN-sequence $\{a_i\}$ generated by the first register R_1 (see Sect. 2.1.5).

Now, for the first companion sequence, let us denote the interleaved PN-sequences by $\left\{v_{d_0^1+i}\right\}, \left\{v_{d_1^1+i}\right\}, \left\{v_{d_2^1+i}\right\}, \ldots, \left\{v_{d_{2^{L_1-1}-1}^1+i}\right\}, i = 0, 1, \ldots,$. We can compute these new positions using Rule 102 and the definition the Zech logarithm as follows:

$$d_k^1 = \mathscr{L}_\alpha\left(d_k^0 - d_{k+1}^0\right) + d_{k+1}^0, \quad k = 0, 1, \ldots, 2^{L_1-1} - 2,$$

$$d_{2^{L_1-1}-1}^1 = \mathscr{L}_\alpha\left(d_{2^{L_1-1}-1}^0 - 1\right) + 1.$$

Similarly, we can compute the shift positions for the jth companion sequence, $j = 1, 2, \ldots, L - 1$ as

$$d_k^j = \mathscr{L}_\alpha\left(d_k^{j-1} - d_{k+1}^{j-1}\right) + d_{k+1}^{j-1}, \quad k = 0, 1, \ldots, 2^{L_1-1} - 2,$$

$$d_{2^{L_1-1}-1}^j = \mathscr{L}_\alpha\left(d_{2^{L_1-1}-1}^{j-1} - 1\right) + 1.$$

Recall that the sequence in the column $t \cdot 2^{L_1-1} + m$, for $m = 0, 1, 2, \ldots,$ $2^{L_1-1} - 1$ and $t = 0, 1, \ldots, L/(2^{L_1-1}) - 1$, is a shifted version of the sequence in the mth column starting in position $t \cdot D \cdot 2^{L_1-1}$. Therefore, we have that:

$$d_k^{t \cdot 2^{L_1-1}+m} = d_k^m + t \cdot D \bmod (2^{L_2} - 1)$$

for $k = 0, 1, \ldots, 2^{L_1-1} - 1$, $m = 0, 1, \ldots, 2^{L_1-1} - 1$ and $t = 0, 1, \ldots, L/(2^{L_1-1}) - 1$.

This means that the positions d_k^j for the companion sequence in the jth column with $j \geq 2^{L_1-1}$ can be computed easily using the positions d_i^s, with $0 \leq s < 2^{L_1-1}$ and without using logarithms.

Example 3.5 Consider again Example 3.4. If we decimate the shrunken sequence and the companion sequence in the 102-CA by distance 2, we obtain that both sequences are composed of interleaving shifted versions of the PN-sequence $\{1\,1\,1\,0\,1\,0\,0\,\ldots\}$ generated by $p(x) = 1 + x^2 + x^3$ (see Table 3.6a and b).

What is more, the positions of both PN-sequences of the shrunken sequence with respect to its first interleaved PN-sequence are $d_0^0 = 0$ and $d_1^0 = 5$, respectively. The positions of the interleaved PN-sequences of the companion sequence with respect to the first PN-interleaved sequence of the shrunken sequence are $d_0^1 = 2$ and $d_1^1 = 4$.

Let us consider the sequence in the 2nd column of the 102-CA. We have seen that this sequence is a shifted version of the shrunken sequence with shift equal to 10. We know that $t = 1$ and $D = 5$, so the two interleaved PN-sequences of this sequence are shifted versions of the first interleaved PN-sequence of the shrunken sequence (Table 3.6c) starting in positions:

$$d_0^2 = d_0^0 + D \cdot 1 \bmod 7 = 5 \quad \text{and} \quad d_1^2 = d_1^0 + D \cdot 1 \bmod 7 = 1, \text{respectively.}$$

Consider again the sequence in the 9th column of the 102-CA. This sequence was a shifted version of the companion sequence, with shift equal to 12. We know that $t = 4$ and $D = 5$, so the two interleaved PN-sequences of this sequence are

Table 3.6 Interleaved PN-sequences of the shrunken sequence and the companion sequences of Example 2.2

	(a)		(b)		(c)		(d)	
	1	0	1	1	(0)	[1]	0	(0)
$d_1^2=1\leftarrow$	[1]	1	0	0	0	1	1	0
$d_0^1=2\leftarrow$	1	0	1	0	1	0	0	1
$d_1^0=d_0^9=3\leftarrow$	0	0	0	1	1	1	0	1
$d_1^1=4\leftarrow$	1	1	0	1	1	0	1	1
$d_0^2=d_0^9=5\leftarrow$	(0)	1	1	1	0	0	1	0
	0	1	1	0	1	1	1	1

the same as the first interleaved PN-sequence of the shrunken sequence (Table 3.6d) starting in positions:

$$d_0^9 = d_0^1 + D \cdot 3 \bmod 7 = 3 \quad \text{and} \quad d_1^9 = d_1^1 + D \cdot 3 \bmod 7 = 5, \text{respectively.}$$

■

3.3.3 Comparison of both Families

In Sect. 3.3.1, we showed that the Fúster–Caballero algorithm produces an hybrid, null 150/90-CA that generates the shrunken sequence. Given two maximal-length LFSRs, this algorithm performs first the Cattell–Muzio algorithm explained in Sect. 3.2.1 and carries out a concatenation procedure to find the CA that generates the shrunken sequence. This fact makes impossible to predict the form of the CA, which depends on L_1 and $p_2(x)$.

In Sect. 3.3.2 since the 102-CAs (60-CAs) are regular we do not need to perform any computations to find the form of the CA; we only need to find its length. On the other hand, according to Theorem 3.4, the length of the 102-CA is, at the most, $T = 2^{L_1-1}(2^{L_2} - 1)$ (when $\gcd(2^{L_2} - 1, T) = 1$), which is greater than $(2^{L_1} - 1)L_2$ (the length of the 150-90-CA given in Sect. 3.3.1). However, the 102-CAs generate 2^{L_1-1} different sequences, the other sequences are shifted versions of these, which is an advantage compared to the 90/150-CA.

As a conclusion, we can say that the 102-CAs are longer but this disadvantage becomes less relevant when we notice the complex process developed in the Fúster–Caballero algorithm to obtain the 150/90-CAs. Besides, this length is reduced to 2^{L_1-1}, since the first 2^{L_1-1} sequences repeat along the 102-CA.

3.4 Modelling the Generalized Self-Shrinking Generator

Since we have seen that the sequences produced by the MSSG and the SSG are sequences produced by the GSSG (Sect. 2.4.3), in this section we only consider the families of CAs that generate the generalized self-shrunken sequences. We recall that the GSS-sequences are a family of sequences generated by a maximum-length LFSR of L stages. We also recall that the characteristic polynomial of the GSS-sequences is of the form $p_t(x) = (1 + x)^t$, with $0 < t \leq 2^{L-1} - (L - 2)$.

3.4.1 Characterization of the 150/90-CA

In this section we present a family of 150/90-CA that generates the family of GSS-sequences.

Table 3.7 GSS-sequences generated by $q(x) = 1 + x^3 + x^4$

	G				S(G)								$p_n(x)$
0	0	0	0	0	0	0	0	0	0	0	0	0	$p_1(x)$
1	0	0	0	1	**0**	**0**	**0**	**1**	**1**	**0**	**1**	**1**	$p_6(x)$
2	0	0	1	0	0	0	1	1	1	1	0	0	$p_5(x)$
3	0	0	1	1	0	0	1	0	0	1	1	1	$p_6(x)$
4	0	1	0	0	0	1	1	1	0	0	1	0	$p_6(x)$
5	0	1	0	1	0	1	1	0	1	0	0	1	$p_5(x)$
6	0	1	1	0	0	1	0	0	1	1	1	0	$p_6(x)$
7	0	1	1	1	0	1	0	1	0	1	0	1	$p_2(x)$
8	1	0	0	0	1	1	1	1	1	1	1	1	$p_1(x)$
9	1	0	0	1	1	1	1	0	0	1	0	0	$p_6(x)$
10	1	0	1	0	1	1	0	0	0	0	1	1	$p_5(x)$
11	1	0	1	1	1	1	0	1	1	0	0	0	$p_6(x)$
12	1	1	0	0	1	0	0	0	1	1	0	1	$p_6(x)$
13	1	1	0	1	1	0	0	1	0	1	1	0	$p_5(x)$
14	1	1	1	0	1	0	1	1	0	0	0	1	$p_6(x)$
15	1	1	1	1	1	0	1	0	1	0	1	0	$p_2(x)$

Theorem 3.5 ([30]) *Given a generalized self-shrunken sequence of period 2^t, $0 \leq t \leq 2^{L-1}$, there exists an hybrid, null 150/90-CA of length 2^t that generates such sequence. Furthermore, the CA will have the form*

$$[90 \quad 150 \quad 150 \quad \ldots \quad 150 \quad 150 \quad 90].$$

Example 3.6 Given a primitive polynomial $p(x) = x^4 + x^3 + 1 \in \mathbb{F}_2[x]$ and an initial state $\{1\,1\,1\,1\}$, the PN-sequence generated is $\{1\,1\,1\,1\,0\,1\,0\,1\,1\,0\,0\,1\,0\,0\,0\ldots\}$. In Table 3.7, it is possible to see the 16 GSS-sequences generated by this PN-sequence. We choose, for example, the sequence number corresponding to $G = 1$, $\{0\,0\,0\,1\,1\,0\,1\,1 \ldots\}$. This sequence has period 8; therefore, there exists a 105/90-CA with length 8 and form

$$[90 \quad 150 \quad 150 \quad 150 \quad 150 \quad 150 \quad 150 \quad 90]$$

that generates such a sequence (see Table 3.8a).   ∎

3.4.2 Characterization of the 102-CA

We start this section with two minor results, whose proofs can be found in [8].

Table 3.8 CAs that generate the GSS-sequence of Example 3.6

(a) 150/90-CA								(b) 102-CA						(c) 60-CA					
90	150	150	150	150	150	150	90	102	102	102	102	102	102	60	60	60	60	60	60
0	0	0	1	0	0	0	1	0	0	0	1	1	1	1	1	1	0	0	0
0	0	1	1	1	0	1	0	0	0	1	0	0	1	1	0	0	1	0	0
0	1	0	1	0	0	1	1	0	1	1	0	1	1	1	1	0	1	1	0
1	1	0	1	1	1	0	1	1	0	1	1	0	1	1	0	1	1	0	1
1	0	0	0	1	0	0	0	1	1	0	1	1	1	1	1	1	0	1	1
0	1	0	1	1	1	0	0	0	1	1	0	0	1	1	0	0	1	1	0
1	1	0	0	1	0	1	0	1	0	1	0	1	1	1	1	0	1	0	1
1	0	1	1	1	0	1	1	1	1	1	1	0	1	1	0	1	1	1	1

Lemma 3.1 ([8, Lemma 2]) *Let $\{u_i\}$ be a binary sequence whose characteristic polynomial is $(x + 1)q(x) \in \mathbb{F}_2[x]$. Then, $q(x)$ generates the sequence $\{v_i\}$, where $v_i = u_i + u_{i+1}$.*

Lemma 3.2 ([8, Theorem 1]) *Let $\{u_i\}$ be a binary sequence whose characteristic polynomial is $p_t(x)$. Then, the characteristic polynomial of the sequence $\{v_i\}$, where $v_i = u_i + u_{i+1}$, is $p_{t-1}(x)$.*

Due to the previous results, we can introduce the following theorem that gives us the length of the CAs that generate the GSS-sequences.

Theorem 3.6 ([10, Theorem 6]) *Given a GSS-sequence with characteristic polynomial $p_t(x)$, there exists a regular, null 102-CA of length t that generates such sequence.*

Recall that the previous results are similar for rule 60. In this case, the 60-CA provides the same sequences but obtained in reverse order. Let us see an illustrative example.

Example 3.7 Consider the GSS-sequence corresponding to $G = 1$ generated by an LFSR with characteristic polynomial $p(x) = 1 + x^3 + x^4$ in Example 3.6:

$$\{0\ 0\ 0\ 1\ 1\ 0\ 0\ 1\ 1\ \ldots\}.$$

According to Theorem 3.6, there exists a regular, null 102-CA of length 6 that generates this sequence as one of its output (vertical) sequences (see Table 3.8b). The characteristic polynomial of this sequence is $p_6(x)$ and thus its linear complexity is $LC = 6$. It is possible to check that the characteristic polynomials of the remaining sequences in the CA are $p_5(x)$, $p_4(x)$, $p_3(x)$, $p_2(x)$ and $p_1(x)$, respectively (consequence of Lemma 3.2). This means that the linear complexity of the (vertical) sequences generated by the null 102-CA is strictly decreasing.

Recall that there exists a 60-CA of length 6 that generates the same exact sequences in inverse order (see Table 3.8c). Therefore, all the results here obtained can be applied to the 60-CA model. ∎

Additionally, as a consequence of Lemmas 3.1 and 3.2, these CAs have a well-defined structure which is given in the following result.

Theorem 3.7 ([10, Theorem 7]) *Consider a GSS-sequence with linear complexity LC. The regular, null 102-CA that generates such a sequence also produces:*

- *The identically 1 sequence (with period 1) in the rightmost column,*
- 2^{i-1} *sequences of period 2^i, for $1 \leq i \leq L - 2$ and*
- $LC - 2^{L-2}$ *sequences of period 2^{L-1} (including the considered GSS-sequence).*

For example, in Table 3.8b we have a 102-CA of length 6 that generates six (vertical) sequences: the identically 1 sequence, one sequence with period 2, two sequences with period 4 and finally, two sequences with period 8 (including the given GSS-sequence).

Comparing the 90/150-CAs given in Sect. 3.4.1 with the 102-CAs (60-CAs) proposed in this section, it can be stated that:

1. Both proposals provide CAs with a defined structure. For the 90/150-CAs, Rule 90 is applied to the extreme cells, while Rule 150 is applied to the remaining cells:

$$[90\ 150\ 150\ \ldots\ 150\ 150\ 90].$$

The 102-CAs are regular; therefore, the same rule is applied for all the cells and the form of the CAs is very simple:

$$[102\ 102\ \ldots\ 102\ 102].$$

2. The length of the proposed 90/150-CAs is 2^{L-1}. On the other hand, the length of the 102-CAs (60-CAs) is the linear complexity of the GSS-sequence considered. We claimed, without proving, that $2^{L-2} < LC \leq 2^{L-1} - (L - 2)$. Therefore, the improvement on the length is not much significant.
3. Finally, in the 90/150-CA model, all the cells (except extreme cells) use Rule 150, which involves the addition of three bits, while the 102-CA (60-CA) model involves the addition of only two bits. Consequently, the number of logic operations to compute the GSS-sequence is much smaller. Furthermore, the periods of the (vertical) sequences of the 102-CA are well known (Theorem 3.7). Therefore, we do not need to compute the whole sequences to complete the CA.

Chapter 4
Cryptanalysis

Cryptanalytic attacks against cryptosystems can be divided into two different classes: direct and indirect attacks.

The direct class attacks the algorithmic nature of the cryptosystem regardless of its implementation.

The indirect class makes use of a physical implementation of the cryptosystem and applies a wide variety of techniques either to give the attacker some intrinsic information about the cryptosystem or to influence its internal state.

In the literature we can find both types of attacks against stream ciphers in general and against the family of shrinking generators in particular, e.g., correlation attacks [40, 50, 90], fast correlation attacks [38, 57, 106], distinguishing attacks [23] or fault attacks [42]. At any rate, this chapter focuses exclusively on the first type of cryptanalytic attacks.

In next sections, we introduce and describe different attacks against the shrinking generator and the self-shrinking generator. Although any correlation attacks are also detailed, most of the cryptanalytic attacks in this chapter make use of the properties of irregular decimation-based generators developed in the two previous chapters. In fact, these attacks exploit:

- the inherent linearity of the output sequences, and
- the modelling of such generators by means of linear CAs.

Successive performance comparisons for the distinct proposals are also provided.

4.1 An Algebraic Attack: Strategic Bits of the Shrinking Generator

This section focuses on an algebraic attack against the shrinking generator [29]. This approach requires less intercepted bits than other attacks. The kernel of this cryptanalysis is the fact that the shrunken sequence is an interleaved sequence, see

Sect. 2.1.2. The properties of the interleaved sequences reveal weaknesses that lead to practical attacks.

In this algebraic attack, the shrinking generator is considered under the following assumptions:

- The key of the cryptosystem is the initial state of both registers.
- The characteristic polynomials of both registers are known and there are no constraints on the number nor the position of their corresponding feedback taps (LFSR's stages included in the feedback loop).
- The intercepted bits are the $(L_2 \times L_1)$ strategic bits of the shrunken sequence.

In this section, we keep the same notation for the shrinking generator as that of Sect. 2.1. In fact, the selector register R_1 has length L_1, characteristic polynomial $p_1(x) \in \mathbb{F}_2[x]$ and its output sequence is denoted by $\{a_i\}$, $i = 0, 1, 2, \dots$. The decimated register R_2 has length L_2, characteristic polynomial $p_2(x) \in \mathbb{F}_2[x]$ and its output sequence is denoted by $\{b_i\}$, $i = 0, 1, 2, \dots$. In addition, the lengths of both registers L_1, L_2 are relatively prime $gcd(L_1, L_2) = 1$ with $L_1 < L_2$, the characteristic polynomials $p_1(x)$, $p_2(x)$ are primitive polynomials in $\mathbb{F}_2[x]$ and both output sequences $\{a_i\}$ and $\{b_i\}$ are PN-sequences of period $T_1 = (2^{L_1} - 1)$ and $T_2 = (2^{L_2} - 1)$, respectively.

The output sequence of the shrinking generator, the shrunken sequence denoted by $\{s_j\}$ $j = 0, 1, 2, \dots$, is a subsequence of $\{b_i\}$ whose terms are chosen according to the positions of the ones (1s) in the sequence $\{a_i\}$. As it was stated in Sect. 2.1, the period T of the shrunken sequence is

$$T = 2^{L_1-1}(2^{L_2} - 1),$$

its linear complexity LC satisfies the inequality

$$L_2 \, 2^{L_1-2} < LC \le L_2 \, 2^{L_1-1},$$

and its characteristic polynomial, $P_{ss}(x)$, is of the form

$$P_{ss}(x) = p(x)^m,$$

where $p(x)$ is a primitive polynomial of degree L_2 (see Theorem 2.2) and m is an integer in the interval $2^{L_1-2} < m \le 2^{L_1-1}$. Due to the period, linear complexity and good statistical properties of the generated sequence, this scheme has been traditionally used as keystream sequence generator in secret-key cryptography.

4.1.1 The Shrunken Sequence as an Interleaved Sequence

In order to cryptanalyse this keystream sequence generator, the $2^{L_1-1}(2^{L_2} - 1)$ bits of a period of the shrunken sequence $\{s_j\}$ can be arranged into a $(2^{L_2} - 1) \times 2^{(L_1-1)}$ matrix that will be called *interleaved matrix* and denoted by IM. In fact,

$$IM = \begin{pmatrix} s_0 & s_1 & s_2 & \cdots & s_{2^{(L_1-1)}-1} \\ s_{2^{(L_1-1)}} & s_{2^{(L_1-1)}+1} & s_{2^{(L_1-1)}+2} & \cdots & s_{2 \cdot 2^{(L_1-1)}-1} \\ s_{2 \cdot 2^{(L_1-1)}} & s_{2 \cdot 2^{(L_1-1)}+1} & s_{2 \cdot 2^{(L_1-1)}+2} & \cdots & s_{3 \cdot 2^{(L_1-1)}-1} \\ s_{3 \cdot 2^{(L_1-1)}} & s_{3 \cdot 2^{(L_1-1)}+1} & s_{3 \cdot 2^{(L_1-1)}+2} & \cdots & s_{4 \cdot 2^{(L_1-1)}-1} \\ \cdots & \cdots & \cdots & \cdots & \cdots \\ s_{(2^{L_2}-2) \cdot 2^{(L_1-1)}} & s_{(2^{L_2}-2) \cdot 2^{(L_1-1)}+1} & s_{(2^{L_2}-2) \cdot 2^{(L_1-1)}+2} & \cdots & s_{(2^{L_2}-1) \cdot 2^{(L_1-1)}-1} \end{pmatrix}.$$

As $\{s_j\}$ is a subsequence of $\{b_i\}$, each element of the matrix IM can be substituted by its corresponding term of the sequence $\{b_i\}$. Thus, IM can be rewritten as follows:

$$IM = \begin{pmatrix} b_{o0} & b_{o1} & \cdots & b_{o(2^{(L_1-1)}-1)} \\ b_{(2^{L_1}-1)+o0} & b_{(2^{L_1}-1)+o1} & \cdots & b_{(2^{L_1}-1)+o(2^{(L_1-1)}-1)} \\ b_{2 \cdot (2^{L_1}-1)+o0} & b_{2 \cdot (2^{L_1}-1)+o1} & \cdots & b_{2 \cdot (2^{L_1}-1)+o(2^{(L_1-1)}-1)} \\ b_{3 \cdot (2^{L_1}-1)+o0} & b_{3 \cdot (2^{L_1}-1)+o1} & \cdots & b_{3 \cdot (2^{L_1}-1)+o(2^{(L_1-1)}-1)} \\ \cdots & \cdots & \cdots & \cdots \\ b_{(2^{L_2}-2) \cdot (2^{L_1}-1)+o0} & b_{(2^{L_2}-2) \cdot (2^{L_1}-1)+o1} & \cdots & b_{(2^{L_2}-2) \cdot (2^{L_1}-1)+o(2^{(L_1-1)}-1)} \end{pmatrix}, \tag{4.1}$$

where the value of the double subindex oj ($j = 0, 1, \ldots, 2^{(L_1-1)} - 1$) depends on the positions of the 1s in the sequence $\{a_i\}$. Indeed, if $a_k = 1$ is the $(j + 1)$th 1 of $\{a_i\}$, then the corresponding subindex $oj = k$.

Recall that j denotes the column of IM and that all the subindices in IM are taken mod $(2^{L_2} - 1)$, that is, the period of the sequence $\{b_i\}$. The number of 1s in the PN-sequence $\{a_i\}$ is 2^{L_1-1}, which is also the number of columns in the matrix IM. Thus, all the elements of any column of IM come from the same term $a_i = 1$ as well as two consecutive elements of any column of IM are two terms of $\{b_i\}$ at a distance $2^{L_1} - 1$.

As different pairs of R_1 and R_2 initial states can generate the same shrunken sequence, in the sequel we assume without loss of generality that the first term of the sequence $\{a_i\}$ equals 1, that is, $a_0 = 1$. Thus, the subindex $o0 = 0$.

Next, the following result analyses the characteristics of the columns of the matrix IM.

Theorem 4.1 ([29]) *The sequences $\{u_j\} = \{b_{k+oj} : k = 0, (2^{L_1} - 1), 2 \cdot (2^{L_1} - 1), \ldots, (2^{L_2} - 2) \cdot (2^{L_1} - 1)\}$ ($j = 0, 1, \ldots, 2^{(L_1-1)} - 1$) corresponding to the columns of the matrix IM are shifted versions of a unique PN-sequence whose characteristic polynomial of degree L_2 is given by*

$$p(x) = (x + \alpha^{T_1})(x + \alpha^{2 \cdot T_1})(x + \alpha^{2^2 \cdot T_1}) \ldots (x + \alpha^{2^{(L_2-1)} \cdot T_1}), \tag{4.2}$$

where $T_1 = 2^{L_1} - 1$ is the period of the PN-sequence $\{a_i\}$ and $\alpha \in \mathbb{F}_{2^{L_2}}$ is a root of the primitive polynomial $p_2(x)$.

The target of this cryptanalysis is the calculation of the initial states of both registers R_1 and R_2. From some known bits of the shrunken sequence, we have to determine the first L_2 bits $\{b_0, b_1, \ldots, b_{L_2-1}\}$ of the PN-sequence $\{b_i\}$ (initial state of R_2) as well as the first L_1 bits $\{a_0, a_1, \ldots, a_{L_1-1}\}$ of the sequence $\{a_i\}$ (initial state of R_1). The number of bits needed for the attack is at most $(L_2 \times L_1)$ bits, which is a negligible value compared with its linear complexity LC. Nevertheless, these bits must be located at very specific positions inside the shrunken sequence. In fact, the required bits (*strategic bits*) are exclusively those terms located at the top-left corner $(L_2 \times L_1)$ submatrix of IM, denoted by SUB_{IM}

$$SUB_{IM} = \begin{pmatrix} b_0 & b_{o1} & \cdots & b_{o(L_1-1)} \\ b_{2^{L_1}-1} & b_{(2^{L_1}-1)+o1} & \cdots & b_{(2^{L_1}-1)+o(L_1-1)} \\ b_{2 \cdot (2^{L_1}-1)} & b_{2 \cdot (2^{L_1}-1)+o1} & \cdots & b_{2 \cdot (2^{L_1}-1)+o(L_1-1)} \\ b_{3 \cdot (2^{L_1}-1)} & b_{3 \cdot (2^{L_1}-1)+o1} & \cdots & b_{3 \cdot (2^{L_1}-1)+o(L_1-1)} \\ \cdots & \cdots & \cdots & \cdots \\ b_{(L_2-1) \cdot (2^{L_1}-1)} & b_{(L_2-1) \cdot (2^{L_1}-1)+o1} & \cdots & b_{(L_2-1) \cdot (2^{L_1}-1)+o(L_1-1)} \end{pmatrix}.$$

It is worth noticing that the strategic bits are not all consecutive. Moreover, between two consecutive rows of the submatrix there is a great number of shrunken sequence bits (as many as $2^{(L_1-1)} - L_1$) whose knowledge is not necessary. The generation of the strategic bits is directly related with the state succession in both registers. Indeed, the first bit of each row of SUB_{IM} is generated when the register states are:

- the initial state of R_1, and
- a state of R_2 at a distance $2^{L_1} - 1$ from the state that generated the first bit of the previous row.

The procedure is repeated systematically for every row of SUB_{IM}. Clearly, the first row of the submatrix is generated from the initial states of R_1 and R_2. Keeping in mind all these features, the cryptanalytic attack is divided into two different steps. In the first one, the computation of the initial state of R_2 is performed. In the second step and based on the R_2 initial state, the corresponding initial state of the register R_1 is determined.

4.1.2 Computation of the R_2 Initial State

The computation of the R_2 initial state is described as follows.

Theorem 4.1 reveals that the column $\{u_0\}$ of IM is the PN-sequence we obtain decimating the sequence $\{b_i\}$ by distance $2^{L_1} - 1$ starting in b_0. Therefore, the bits $\{b_0, b_1, \ldots, b_{L_2-1}\}$ are terms of $\{u_0\}$. For the sake of simplicity such a column will be denoted by $\{u_0\} = \{u_i\}$ $(i = 0, 1, \ldots, 2^{L_2} - 2)$. In addition, the first column of the submatrix SUB_{IM} corresponds to the first L_2 bits $\{u_0, u_1, \ldots, u_{L_2-1}\}$ of $\{u_0\}$, which are known.

According to the properties of the PN-sequences [41], any term u_k of $\{u_i\}$ can be expressed as a linear combination of the first L_2 bits $\{u_0, u_1, \ldots, u_{L_2-1}\}$ by means of the modular expression:

$$q(x) = x^k \bmod p(x),$$

where $q(x) = c_{L_2-1}x^{L_2-1} + \ldots + c_1 x + c_0 \in \mathbb{F}_2[x]$ and $c_i \in \mathbb{F}_2$.

Thus, the term u_k can be written as:

$$u_k = c_{L_2-1}u_{L_2-1} + \ldots + c_1 u_1 + c_0 u_0. \tag{4.3}$$

Therefore, the computation of the bits $\{b_0, b_1, \ldots, b_{L_2-1}\}$ is reduced to:

1. Determine the positions of the terms $b_0, b_1, \ldots, b_{L_2-1}$ in the sequence $\{u_i\}$.
2. Compute the value of such terms.

From Eq. (4.1), it is clear that $b_{n \cdot (2^{L_1}-1)}$ is the $(n + 1)$th element of the first column of IM. Thus, solving the following system of modular equations in the unknowns n_i:

$$\left\{ n_i \cdot (2^{L_1} - 1) \equiv i \mod 2^{L_2} - 1 \qquad (i = 0, 1, \ldots, L_2 - 1), \right. \tag{4.4}$$

we can determine the positions of the R_2 initial state bits in the sequence $\{u_i\}$. Next making use of Eq. (4.3), we compute the values of $\{u_{n_i}\}$ $(i = 0, 1, \ldots, L_2 - 1)$ that correspond to the bits $u_{n_i} = b_i$. Consequently, the initial state of R_2, $\{b_0, b_1, \ldots, b_{L_2-1}\}$, has been uniquely determined.

4.1.3   Computation of the R_1 Initial State

The computation of the R_1 initial state is described as follows.

As before, from the first column of SUB_{IM} and Eq. (4.3), any term u_k of $\{u_i\}$ can be calculated. Over the sequence $\{u_0\}$ we compute $(L_2 - 1)$ subsequences of L_2 consecutive bits, denoted by $\{B_i\}$ $(i = 1, 2, \ldots, L_2 - 1)$, starting each of them in the term u_{n_i} $(i = 1, 2, \ldots, L_2 - 1)$, respectively. That is $\{B_i\} = \{u_{n_i}, u_{n_i+1}, u_{n_i+2}, \ldots, u_{n_i+L_2-1}\}$ where the integers n_i are defined in (4.4).

Since the columns $\{u_j\}$ $(j = 0, 1, \ldots, 2^{(L_1-1)} - 1)$ of IM are exactly the same PN-sequence but starting at different points, we try to determine the first element of each $\{u_j\}$ $(j = 1, \ldots, L_1 - 1)$ as a term of $\{b_i\}$. Therefore, we compare each subsequence $\{B_i\}$ with the successive columns of the submatrix SUB_{IM} in order to check the coincidence of the L_2 bits compared. More precisely, we compare $\{B_i\}$ $(i = 1, 2, \ldots, L_2 - 1)$ with the first L_2 bits of $\{u_1\}$. If there is coincidence for

$i = k$, then u_{n_k} is the first element of $\{u_1\}$. Therefore, $u_{n_k} = b_k$ and the bits of the R_1 initial state will be: $a_0 = 1, a_1 = a_2 = \ldots = a_{k-1} = 0$ and $a_k = 1$.

Next we compare $\{B_i\}$ $(i = k+1, \ldots, L_2 - 1)$ with the first L_2 bits of $\{u_2\}$. If there is coincidence for $i = l$, then u_{n_l} is the first element of $\{u_2\}$. Therefore, $u_{n_l} = b_l$ and the bits of the R_1 initial state will be: $a_0 = 1, a_1 = a_2 = \ldots = a_{k-1} = 0$, $a_k = 1, a_{k+1} = a_{k+2} = \ldots = a_{k+l-1} = 0, a_{l+k} = 1$ and so on. The procedure ends when the last bit a_{L_1-1} has been determined. The key idea is to identify the first element of each $\{u_j\}$ with its corresponding term of the sequence $\{b_i\}$. Then we deduce easily the bits $\{a_0, a_1, \ldots, a_{L_1-1}\}$.

In brief, the simple comparison of the subsequences $\{B_i\}$ with the successive columns of SUB_{IM} allows one to compute the R_1 initial state $\{a_0, a_1, \ldots, a_{L_1-1}\}$.

An illustrative example clarifies this cryptanalysis.

4.1.4 An Illustrative Example

Let us consider a shrinking generator characterized by:

1. R_2 with length $L_2 = 5$, characteristic polynomial $p_2(x) = 1 + x^2 + x^3 + x^4 + x^5$ and output sequence $\{b_i\}$.
2. R_1 with length $L_1 = 4$, characteristic polynomial $p_1(x) = 1 + x^3 + x^4$ and output sequence $\{a_i\}$.
3. The characteristic polynomial of the shrunken sequence is $P_{ss}(x) = p(x)^m = (1 + x + x^2 + x^3 + x^5)^8$.

Given 20 bits of the shrunken sequence corresponding to the (5×4) submatrix SUB_{IM}

$$SUB_{IM} = \begin{pmatrix} 1\,0\,1\,1 \\ 1\,0\,0\,1 \\ 0\,1\,0\,1 \\ 0\,1\,1\,1 \\ 0\,0\,0\,1 \end{pmatrix},$$

we can launch a cryptanalytic attack against the shrinking generator to obtain the initial states of both registers.

Computation of the R_2 Initial State According to Sect. 4.1.2, we compute the positions of $\{b_0, b_1, \ldots, b_4\}$ in $\{u_0\}$ solving the system of modular equations:

$$\left\{ \begin{array}{ll} n_i \cdot 15 \equiv i \mod 31 & (i = 0, 1, \ldots, 4). \end{array} \right.$$

That is $n_0 = 0, n_1 = 29, n_2 = 27, n_3 = 25$ and $n_4 = 23$. According to the first column of SUB_{IM}, we have $u_0 = 1, u_1 = 1$ and $u_2 = u_3 = u_4 = 0$. Then, via

Eq. (4.3), we write:

$$u_{n_0} = u_0 = 1$$

$$u_{n_1} = u_{29} = u_4 + u_3 + u_2 = 0$$

$$u_{n_2} = u_{27} = u_2 + u_1 + u_0 = 0$$

$$u_{n_3} = u_{25} = u_3 + u_1 = 1$$

$$u_{n_4} = u_{23} = u_4 + u_2 + u_0 = 1.$$

As $u_{n_i} = b_i$ ($i = 0, 1, \ldots, 4$), then the initial state of the register R_2 is $\{b_0, b_1, b_2, b_3, b_4\} = \{1, 0, 0, 1, 1\}$.

Table 4.1 shows the calculations performed to cryptanalyse this shrinking generator. In fact, the most left column in the table represents the generic index k numbered $(0, 1, \ldots, 2^{L_2} - 2 = 30)$. Next column shows the positions n_i ($i = 0, 1, \ldots, 4$) of the terms $\{b_0, b_1, \ldots, b_4\}$ of the sequence $\{b_i\}$. The following columns of Table 4.1 represent the matrix IM: in boldface the top-left corner (5×4) submatrix SUB_{IM} with the known bits and the symbol $-$ corresponds to unknown bits of the shrunken sequence not needed for the cryptanalysis.

Computation of the R_1 Initial State According to Sect. 4.1.3, we compute 4 subsequences $\{B_i\}$ ($i = 1, 2, \ldots, 4$) of 5 consecutive bits where

$$\{B_i\} = \{u_{n_i}, u_{n_i+1}, u_{n_i+2}, u_{n_i+3}, u_{n_i+4}\} \qquad (i = 1, 2, 3, 4).$$

It is worth noticing that the bits of $\{u_0\}$ used in $\{B_i\}$ are all concentrated in the last positions of the sequence, see Corollary 2.2. Thus, the application of the linear recurrence relationship given by the polynomial $p(x)$, that is, $u_{n+5} = u_{n+3} + u_{n+2} + u_{n+1} + u_n$ with ($n = 30, 29, 28, \ldots, 23$), is here enough to compute the terms used in $\{B_i\}$, see Table 4.1.

Therefore, $\{B_1\} = \{0, 0, 1, 1, 0\}$, $\{B_2\} = \{0, 1, 0, 0, 1\}$, $\{B_3\} = \{1, 0, 0, 1, 0\}$ and $\{B_4\} = \{1, 1, 1, 0, 0\}$.

Now we compare $\{B_1\} = \{0, 0, 1, 1, 0\}$ with the first 5 bits of $\{u_1\}$. There is coincidence for $i = 1$, then $u_{n_1} = u_{29}$ is the first element of $\{u_1\}$. Therefore, $u_{29} = u_{n_1} = b_1$ and the bits of the R_1 initial state will be: $a_0 = 1$ and $a_1 = 1$.

Next we compare $\{B_2\} = \{0, 1, 0, 0, 1\}$ with the first 5 bits of $\{u_2\}$. There is no coincidence.

Then we compare $\{B_3\} = \{1, 0, 0, 1, 0\}$ with the first 5 bits of $\{u_2\}$. There is coincidence for $i = 3$, then $u_{n_3} = u_{25}$ is the first element of $\{u_2\}$. Therefore, $u_{25} = u_{n_3} = b_3$ and the bits of the R_1 initial state will be: $a_0 = 1$, $a_1 = 1$, $a_2 = 0$ and $a_3 = 1$.

The procedure ends as the last bit a_3 has been determined. Therefore, the R_1 initial state is $\{a_0, a_1, a_2, a_3\} = \{1, 1, 0, 1\}$.

Table 4.1 Matrix IM corresponding to the described shrinking generator

k	$\{b_i\}$	$\{u_0\}$	$\{u_1\}$	$\{u_2\}$	$\{u_3\}$	$\{u_4\}$	$\{u_5\}$	$\{u_6\}$	$\{u_7\}$
0	b_0	1	0	1	1	–	–	–	–
1		1	0	0	1	–	–	–	–
2		0	1	0	1	–	–	–	–
3		0	1	1	1	–	–	–	–
4		0	0	0	1	–	–	–	–
5		–	–	–	–	–	–	–	–
6		–	–	–	–	–	–	–	–
7		–	–	–	–	–	–	–	–
8		–	–	–	–	–	–	–	–
9		–	–	–	–	–	–	–	–
10		–	–	–	–	–	–	–	–
11		–	–	–	–	–	–	–	–
12		–	–	–	–	–	–	–	–
13		–	–	–	–	–	–	–	–
14		–	–	–	–	–	–	–	–
15		–	–	–	–	–	–	–	–
16		–	–	–	–	–	–	–	–
17		–	–	–	–	–	–	–	–
18		–	–	–	–	–	–	–	–
19		–	–	–	–	–	–	–	–
20		–	–	–	–	–	–	–	–
21		–	–	–	–	–	–	–	–
22		–	–	–	–	–	–	–	–
23	b_4	1	–	–	–	–	–	–	–
24		1	–	–	–	–	–	–	–
25	b_3	1	–	–	–	–	–	–	–
26		0	–	–	–	–	–	–	–
27	b_2	0	–	–	–	–	–	–	–
28		1	–	–	–	–	–	–	–
29	b_1	0	–	–	–	–	–	–	–
30		0	–	–	–	–	–	–	–

Notice that just the knowledge of three columns of the submatrix SUB_{IM} has been necessary to identify the initial state of R_1. Indeed, the number of columns needed equals the number of 1s in the initial state of the selector register. The maximum number of needed bits corresponds to $L_2 \times L_1$ when the R_1 initial state is the identically 1 state. In the remaining cases, less bits are sufficient.

Once the initial states of both register have been determined, the whole shrunken sequence, that is, the keystream sequence, can be computed.

The previous cryptanalysis shows that not all the bits in the shrunken sequence have the same relevance. Some bits are much more important than others. Thus, the knowledge of a few strategic bits allows one to launch a simple and efficient attack whose computational complexity is minimum $O(L_2)$. In fact, the number of intercepted bits is at most $L_2 \times L_1$, while the computational requirements are reduced to solve a system of L_2 modular equations and successive applications (no more than L_2^2) of a linear recurrence relationship.

In brief, if the attacker without knowledge of the key obtains the physical encryption device and can manipulate it to get the strategic bits, then this simple approach is realistic.

4.2 Linear Consistency Test Against the Shrinking Generator

In this section, a cryptanalysis of the shrinking generator based on the linear consistency test (LCT) [104] is introduced.

In keystream sequence generators, it is sometimes possible to single out a certain subkey K_1 from the entire cryptosystem key K and write out a system of linear equations:

$$Ax = b, \tag{4.5}$$

where the coefficient matrix A is determined by the own keystream generator and parameterized by K_1, the fixed right-side vector b is the segment of intercepted bits and the solution vector x can be used to compute the remaining bits of the key K. If the proposed K_1 coincides with the right subkey used in generating the intercepted sequence, then such a system of linear equations will certainly be consistent. Otherwise, the consistency probability will be very small [104, Theorem 1]. Thus, in order to find the right subkey K_1, we only need to test the consistency of the linear equation system with respect to all possible choices of the subkey K_1. In this way, the amount of work needed is dramatically reduced regarding the exhaustive search of the entire key K.

The cryptanalysis here presented combines the LCT method with the fact that the shrunken sequence is an interleaved sequence (see Sects. 2.1 and 4.1.1). More precisely, this cryptanalytic attack applies the LCT to the interleaved PN-sequences of the shrunken sequence [12].

In this algebraic attack, the shrinking generator is considered under the following assumptions:

- The key of the cryptosystem is the initial states of both registers.
- The characteristic polynomials of both registers are known and there are no constraints on the number nor the position of their corresponding feedback taps.
- The intercepted bits used in this attack are N consecutive bits of the shrunken sequence.

In this section, we keep the same notation for the shrinking generator as that of Sects. 2.1 and 4.1. Indeed, the selector register R_1 has length L_1, characteristic polynomial $p_1(x) \in \mathbb{F}_2[x]$ and its output sequence is denoted by $\{a_i\}$, $i = 0, 1, 2, \ldots$. The decimated register R_2 has length L_2, characteristic polynomial $p_2(x) \in \mathbb{F}_2[x]$ and its output sequence is denoted by $\{b_i\}$, $i = 0, 1, 2, \ldots$. Moreover, the lengths of both registers L_1, L_2 are relatively prime $gcd(L_1, L_2) = 1$ with $L_1 < L_2$, the characteristic polynomials $p_1(x)$, $p_2(x)$ are primitive polynomials in $\mathbb{F}_2[x]$ and both output sequences $\{a_i\}$ and $\{b_i\}$ are PN-sequences of period $T_1 = (2^{L_1} - 1)$ and $T_2 = (2^{L_2} - 1)$, respectively. As before, the output sequence of the shrinking generator, the shrunken sequence, is denoted by $\{s_j\}$ $j = 0, 1, 2, \ldots$.

Previously to the attack description, additional notation is introduced:

- The initial states of registers R_1 and R_2 are denoted by $is_1 = \{a_0, a_1, \ldots, a_{L_1-1}\}$ and $is_2 = \{b_0, b_1, \ldots, b_{L_2-1}\}$, respectively.
- The subsequence $\mathbf{S} = \{s_0, s_1, \ldots, s_{N-1}\}$ denotes the N intercepted bits of the shrunken sequence. Currently, the number N can be written as $N = N_1 + N_2$, where N_1 bits are used to compute the pair (is_1, is_2), while N_2 bits are used to check the correctness of the previous pair.
- According to Proposition 2.1, δ is an integer $\delta \in \{1, 2, 3, \ldots, T_2 - 1\}$, such that $T_1 \delta \equiv 1 \bmod T_2$.

As it was seen in Sect. 4.1.1, the shrunken sequence can be written as a $(2^{L_2} - 1) \times 2^{(L_1-1)}$ matrix IM whose columns, the sequences $\{\mathbf{u}_j\} = \{b_{k+oj} : k = 0, (2^{L_1} - 1), 2 \cdot (2^{L_1} - 1), \ldots, (2^{L_2} - 2) \cdot (2^{L_1} - 1)\}$ ($j = 0, 1, \ldots, 2^{(L_1-1)} - 1$), are shifted versions of a unique PN-sequence with characteristic polynomial $p(x)$ defined in (4.2). Now the N intercepted bits are elements of the successive interleaved PN-sequence $\{\mathbf{u}_j\}$. Nevertheless, in this attack we only focus on the first interleaved sequence $\{\mathbf{u}_0\}$. For the sake of simplicity such a column will be denoted by $\{\mathbf{u}_0\} = \{u_i\}$ ($i = 0, 1, \ldots, 2^{L_2} - 2$). As in Sect. 4.1.2, any term u_k of $\{u_i\}$ can be expressed as a linear combination of the first L_2 bits $\{u_0, u_1, \ldots, u_{L_2-1}\}$ by means of the modular expression:

$$q(x) = x^k \bmod p(x),$$

where $q(x) = c_{L_2-1} x^{L_2-1} + \ldots + c_1 x + c_0 \in \mathbb{F}_2[x]$ and $c_i \in \mathbb{F}_2$.

Thus, the term u_k can be written as:

$$u_k = c_{L_2-1} u_{L_2-1} + \ldots + c_1 u_1 + c_0 u_0.$$

The proposed attack is divided into two phases:

In phase 1, we check the 2^{L_1-1} initial states is_1 starting with 1 (as only the 1s of $\{a_i\}$ generate bits in the shrunken sequence) to determine a set Q of possible candidates for initial state of R_1. Each checking applies the LCT to a linear equation system.

In phase 2, for every is_1 in Q its corresponding is_2 will be computed. The pair (is_1, is_2) able to generate the shrunken sequence will be the key of the cryptosystem. Next, a general outline of the algorithm is depicted.

Algorithm 1 Cryptanalysis of the shrinking generator

Input: Lengths L_1, L_2, polynomials $p_1(x)$, $p_2(x)$ and N intercepted bits $\{s_0, s_1, \ldots s_{N-1}\}$
01:   Compute Phase 1.
02:   Compute Phase 2.

Output:
    The pair (is_1, is_2) that generates the shrunken sequence.

Now a detailed description of the algorithm is given.
Phase 1:
For each is_1 **do:**

Step 1:   Starting in is_1, generate a portion of sequence $\{a_i\}$ until N_1 1s are obtained. Such 1s will be located at positions i_k $(k = 0, 1, \ldots, N_1 - 1)$ in $\{a_i\}$.
Step 2:   Determine N_1 positions in the sequence $\{u_i\}$ as

$$d_k \equiv \delta \cdot i_k \bmod T_2 \qquad (k = 0, 1, \ldots, N_1 - 1).$$

Step 3:   Assign the N_1 intercepted bits to the previous positions

$$u_{d_k} = s_k \qquad (k = 0, 1, \ldots, N_1 - 1).$$

Step 4:   Express each u_{d_k} as a function of the first L_2 terms of $\{u_i\}$, that is, $u_{d_k} = f_k(u_0, u_1, \ldots, u_{L_2-1})$, by means of

$$x^{d_k} \bmod p(x) \qquad (k = 0, 1, \ldots, N_1 - 1).$$

It turns out to be a system of linear equations

$$\left\{ f_k(u_0, u_1, \ldots, u_{L_2-1}) = s_k \qquad (k = 0, 1, \ldots, N_1 - 1), \right. \tag{4.6}$$

with N_1 equations in the $\{u_0, u_1, \ldots, u_{L_2-1}\}$ unknowns.
Step 5:   Apply the Linear Consistency Test (LCT) to check the consistency of the previous system,
    if the system is consistent, **then** include is_1 in Q
    else is_1 is rejected.

end do

The result of this phase is the set Q of possible candidates for initial state of R_1 that will be checked in the next phase. Recall that the state is_1 under consideration parameterizes the coefficient matrix in the system of linear equations (4.6) and that the subkey K_1 is is_1.

Phase 2:

For each is_1 in Q **do**:

Step 1: Express each b_{i_k} as a function of the first L_2 terms of $\{b_i\}$, that is, $b_{i_k} = g_k(b_0, b_1, \ldots, b_{L_2-1})$, by means of

$$x^{i_k} \bmod p_2(x) \qquad (k = 0, 1, \ldots, N_1 - 1).$$

It turns out to be a system of linear equations

$$\left\{ g_k(b_0, b_1, \ldots, b_{L_2-1}) = s_k \qquad (k = 0, 1, \ldots, N_1 - 1), \right.$$

with N_1 equations in the $\{b_0, b_1, \ldots, b_{L_2-1}\}$ unknowns.

Step 2: Apply the Linear Consistency Test (LCT) to check the consistency of the previous system,

if the system is not consistent, **then** reject (is_1, is_2)

else if the pair (is_1, is_2) can generate the shrunken sequence (checked by means of the N_2 bits),

then cryptosystem broken !!!

else is_1 is rejected.

end do

The result of this phase is the pair (is_1, is_2) generating the shrunken sequence that is the key of the cryptosystem.

For illustrative purposes, a software implementation of the previous attack has been performed on a laptop device with the following specifications:

- Operative system: Arch Linux
- CPU: Dual core Intel Core i7-4510U, Cache 4096 KB, Freq. 3100 MHz
- RAM: 8 GB, Type: DDR3
- Hard Disk: Type SSD, Size 256.1 GB

The application of the LCT to this generator writes out a system of linear equations equal to that of Eq. (4.5), where A is an $(N_1 \times L_2)$ binary coefficient matrix, $\mathbf{x}$ is the $(L_2 \times 1)$ vector of unknowns and $\mathbf{b}$ is the $(N_1 \times 1)$ right-side vector of intercepted bits. Each initial state is_1 parameterizes the coefficient matrix A as each is_1 writes out a different system of linear equations (4.6). Then the linear consistency test (LCT) [104] checks the consistency of the corresponding equation system. The running time of the attack is dominated by phase 1 which has a time complexity of $O(2^{L_1-1} \cdot (N_1 \times L_2)^3)$ that is exponential in L_1 due to the number of is_1 considered and polynomial in L_2. In fact, the work factor needed for each test is that of the Gauss elimination algorithm applied to the augmented matrix $(A, \mathbf{b})$, which is cubic in the dimension of the matrix. In any case, the cubic factor is irrelevant compared with the exponential factor.

Some numerical results are depicted in Table 4.2 where L_1, L_2 are the lengths of registers R_1 and R_2, respectively, T is the period of the corresponding shrunken

Table 4.2 Numerical results
for the algorithm

L_1	L_2	T	N_1	$c(Q)$	t (s)
4	5	248	10	1	0.0064
5	6	1008	12	1	0.0173
9	10	261,888	20	1	0.3856
10	11	1,048,064	22	1	0.8552
11	12	4,193,280	24	1	1.8114
12	13	16,775,168	26	1	4.2623
13	14	67,104,768	28	1	9.0739
14	15	268,427,264	30	1	20.0681
15	16	$1.0737 \cdot 10^9$	32	1	44.9963
16	17	$4.2949 \cdot 10^9$	34	2	98.1865
17	18	$1.7180 \cdot 10^{10}$	36	1	217.9489
18	19	$6.8719 \cdot 10^{10}$	38	2	477.1288
19	20	$2.7488 \cdot 10^{11}$	40	1	1092.7125
20	21	$1.0995 \cdot 10^{12}$	42	1	2327.2800
21	22	$4.3980 \cdot 10^{12}$	44	1	4997.0925

sequence, N_1 is the number of intercepted bits for computation, $c(Q)$ is the cardinal of Q, that is, the number of candidates for initial state of R_1, and t is the running time expressed in seconds. It must be noticed that the period of the shrunken sequence is much greater than the number of intercepted bits needed to successfully run the algorithm within a reasonable time. In fact, $N_1 = 2 \cdot L_2$, while N_2 is currently chosen as $N_2 = N_1$. In brief, the requirements of intercepted sequence are extremely low. In Table 4.3, the same results are shown but now the number of intercepted bits N_1 equals L_2. In this case, since N_1 has been reduced, the execution time has been reduced too. Nevertheless, the number of candidates has grown considerably. Table 4.4 shows the numerical results corresponding to the verification of a unique initial state *is* 1 in the phase 1 of the algorithm. Recall that even for large values of L_1 and L_2 the execution time of such routine is very low. The program makes use of SageMath, an algebraic computation systems based on Python. In order to handle polynomials in $\mathbb{F}_2[x]$, SageMath uses the libraries NLT. In order to compute with matrices over $\mathbb{F}_2$, SageMath uses the libraries M4RI. In the LCT application, the system of equations is transformed into a low reduced echelon form. This step is important in the computation efficiency as the system consistency is reduced to test the existence of a row $\{0, 0, \ldots, 0, 1\}$ in the coefficient matrix of the system.

Both phases of this algorithm are fully parallelizable and some tweaks can be made to optimize the LCT step.

The most remarkable features of the proposed attack are:

1. The low amount of intercepted bits needed for its execution. In fact, $N_1 = n \cdot L_2$, n being a small integer ($n = 2, 3, 4$), and $N_2 \leq N_1$ are enough to conclude this attack successfully. Thus the amount of sequence required is linear in the length of the register R_2.

Table 4.3 Numerical results for the algorithm when $N_1 = L_2$

L_1	L_2	T	N_1	$c(Q)$	t (s)
4	5	248	5	5	0.0046
5	6	1008	6	14	0.0099
6	7	4064	7	25	0.0216
7	8	16,320	8	46	0.0513
8	9	65,408	9	78	0.11969
9	10	261,888	10	160	0.2478
10	11	1,048,064	11	210	0.7123
11	12	4,193,280	12	708	1.3290
12	13	16,775,168	13	1183	3.1078
13	14	67,104,768	14	2227	6.0204
14	15	268,427,264	15	4494	13.0011
15	16	$1.0737 \cdot 10^9$	16	8710	29.4033
16	17	$4.2949 \cdot 10^9$	17	6183	57.9891
17	18	$1.7180 \cdot 10^{10}$	18	35351	151.4661

Table 4.4 Numerical results for the verification of one $is1$

L_1	L_2	N_1	t (s)
5	6	12	0.00080
6	7	14	0.00106
7	8	16	0.00112
8	9	18	0.00130
9	10	20	0.00157
10	11	22	0.00169
20	21	42	0.00911
30	31	62	0.01044
40	41	82	0.01980
50	51	102	0.03160
59	60	120	0.03547
60	61	122	0.03794
61	62	124	0.03806
62	63	126	0.04035
63	64	128	0.04108

2. The maximum complexity for attacking such a model is $O(2^{L_1-1} \cdot (N_1 \times L_2)^3)$, exponential in L_1 and polynomial in L_2. Thus, in terms of complexity we can say that the computational complexity of this cryptanalysis is $O(2^{L_1-1})$.
3. The attack is deterministic and always recovers both initial states.

Compared with the cryptanalysis described in Sect. 4.1, we can see that the amount of intercepted sequence is very low in both cases. Nevertheless, as the N bits intercepted from the shrunken sequence are now consecutive, then the computational complexity is much greater.

Another cryptanalysis against the shrinking generator that makes use of the LCT without considering the shrunken sequence as an interleaved sequence can be found in [75]. In this approach, the maximum complexity is $O(2^{L_1})$ and the requirements of intercepted sequence are greater than in the previous cryptanalytic attack.

4.3 102-CA Recovery Attack Against the Shrinking Generator

In this section, we present two deterministic attacks based on 102-CAs (60-CAs) that recover the shrunken sequence.

4.3.1 Overlapping Attack

In order to recover the shrunken sequence, we make use of the properties of the 102-CAs (60-CAs) that generate such a sequence. Mainly, we take advantage of the fact that shifted versions of the shrunken sequence and other companion sequences appear several times along the 102-CA (see Sect. 3.3.2).

In order to illustrate this idea, consider again Example 3.4. We had a shrunken sequence of period $T = 14$ and a 102-CA of length 14 that generated such a sequence in its leftmost column (see Table 4.5). If we intercept the first 6 bits of the shrunken sequence, then we can recover 21 elements in the CA using rule 102 (see the triangle in the top-left corner of the 102-CA in Table 4.5). According to the 102-CA properties, shifted versions of the shrunken sequence and other companion sequences are repeated along the 102-CA (see Example 3.4). Thus, we can recover

Table 4.5 CA that generates the shrunken sequence in Example 3.4

102	102	102	102	102	102	102	102	102	102	102	102	102	102
1	1	0	1	0	0	1	0	0	1			1	1
0	1	1	1	0		1	0	1				0	
1	0	0	1			1	1	1	1	0	1	0	0
1	0	1				0		0	1	1	1	0	
1	1	1	1	0	1	0	0	1	0	0	1		
0		0	1	1	1	0			1	0	1		
0	0	1	0	0	1			1	1	1	1	0	1
0		1	0	1				0		0	1	1	1
		1	1	1	1	0	1	0	0	1	0	0	1
		0		1	1	1	0			1	0	1	
0	1	0	0	1	0	0	1			1	1	1	1
1	1	0		1	0	1				0		0	1
0	1			1	1	1	1	0	1	0	0	1	0
1			0		1	1	1	0			1	0	0

the same number of bits in other positions (see the other triangles in Table 4.5). In this case, the triangles of recovered bits overlap. Therefore, we can recover all the elements in the CA and, consequently, all the shrunken sequence.

In general, we need to intercept

$$N = 2^{L_1-1}(2^{L_2} - \mathcal{Z}_\alpha(1))$$

bits of the shrunken sequence for the triangles to overlap. This number depends on L_1, L_2 and the value of $\mathcal{Z}_\alpha(1)$, which in turn depends on the characteristic polynomial $p(x)$ of the interleaved PN-sequences (see Theorem 2.1). Nevertheless, this number of required bits might be unrealistic for practical applications. For instance, if we consider a shrinking generator with $L_1 = 3$ and $L_2 = 4$, then the period of the shrunken sequence is $T = 60$. There are two primitive polynomials of degree 4, $q_1(x) = 1 + x + x^4$ and $q_2(x) = 1 + x^3 + x^4$, for which $\mathcal{Z}_\alpha(1)$ takes the values 4 and 12, respectively. Thus, we need $N_1 = 48$ intercepted bits for $p(x) = q_1(x)$ and $N_2 = 16$ intercepted bits for $p(x) = q_2(x)$. The designer would have chosen $q_1(x)$, while the cryptanalyst would have preferred $q_2(x)$.

4.3.2 CA-Based Cryptanalysis Against the Shrinking Generator

In this subsection, we introduce a cryptanalysis against the shrinking generator based on the results given in Sects. 2.1 and 3.3.2. The attack [11] performs an exhaustive search over the 2^{L_1-1} initial states of the register R_1 starting with 1.

Given n bits of the shrunken sequence, denoted by $s = \{s_0, s_1, \ldots, s_{n-1}\}$, Algorithm 2 checks the correctness of a given R_1 initial state $a = \{a_0, a_1, \ldots, a_{L_1-1}\}$. For each checked state, we perform rule 102 on the intercepted bits to generate as many bits as possible in the sequences of 102-CA (see Sect. 3.3.2). The idea is to determine bits of the first interleaved PN-sequence of the shrunken sequence by using not only the n intercepted bits but also the bits obtained from other companion sequences. If two different bits are assigned to the same position, then there is a discrepancy and the initial state a is rejected. Otherwise, Algorithm 2 returns the matrix M with the values and positions of the recovered bits in the first interleaved sequence. Then, the R_2 initial state bits are computed via Proposition 2.1.

It is worth pointing out that this attack can be equivalently designed with rule 60, $x_i^{t+1} = x_{i-1}^t + x_i^t$. In that case, the sequences would appear in reverse order along the CA, but the results would be identical. Next a numerical example is introduced.

Example 4.1 Consider a shrinking generator with two registers R_1 and R_2 whose characteristic polynomials are $p_1(x) = 1 + x + x^6$ and $p_2(x) = 1 + x^3 + x^7$, respectively. Now the period of the shrunken sequence is $2^5(2^7 - 1) = 4064$. Since $L_2 = L_1 + 1$, we know that $p(x)$ is the reciprocal polynomial of $p_2(x)$

Algorithm 2 Crypto: Test an initial state for R_1

Input: $p_1(x)$, $p(x)$, δ, $\boldsymbol{s}$ and $\boldsymbol{a}$
function $[M, Stop] = \mathbf{Crypto}(p_1(x), p(x), \delta, \boldsymbol{s}, \boldsymbol{a})$

01: Compute $\{a_i\}$ using $p_1(x)$ and $\boldsymbol{a}$ until finding $n = length(\boldsymbol{s})$ ones;
02: Store in P the positions of the n 1s;
03: Store in P the new positions computed as $P_i \cdot d \mod (2^{L_2} - 1)$;
04: Store $[P_i, s_i]$ in a matrix M;
05: $Stop = 1$ and $\ell = length(P)$;
06: **while** $Stop = 1$ **and** $\ell > 1$ **do**
07:     Update P and ℓ with the new positions;
08:     Update $\boldsymbol{s}$ with $\{s_0 + s_1, s_1 + s_2, \ldots, s_{n-2} + s_{n-1}\}$;
09:     Store [m,n] = size(M);
10:     **for** $j = 0$ **to** $m - 1$ **do**
11:        **for** $k = 0$ **to** $length(P) - 1$ **do**
12:           **if** $M_{j1} = P_k$ **and** $M_{j2} \neq s_k$ **then**
13:              Initialise M with zeros;
14:              $Stop = 0$;
15:           **end if**
16:           Store $[P_k, s_k]$ in M;
17:        **end for**
18:     **end for**
19: **end while**
end function
Output:
M: Recovered bits and their positions in the first interleaved PN-sequence.
$Stop$: 1 if the initial state is considered correct and 0 otherwise.

(Corollary 2.1), then $p(x) = 1 + x^4 + x^7$. Moreover, according to Proposition 2.1, the decimation distance is $\delta = T_2 - 2 = 2^7 - 3 = 125$.

Assume we intercept $n = 6$ bits of the shrunken sequence: $\boldsymbol{s} = \{1, 0, 1, 0, 0, 0\}$.

In order to check the correctness of an R_1 initial state, e.g., $\boldsymbol{a} = \{1, 1, 1, 1, 0, 1\}$, we apply Algorithm 2.

Input: $p_1(x) = 1 + x + x^6$, $p(x) = 1 + x^4 + x^7$, $\boldsymbol{a} = (1, 1, 1, 1, 0, 1)$, $\delta = 125$ and $\boldsymbol{s} = \{1, 0, 1, 0, 0, 0\}$.

Compute the PN-sequence generated by R_1 starting in $\boldsymbol{a}$ until 6 ones are found: $\{1, 1, 1, 1, 0, 1, 0, 0, 0, 1\}$.

The positions of the 1s are: $pos = \{0, 1, 2, 3, 5, 9\}$.

The positions of the 6 intercepted bits in the first interleaved PN-sequence are: $d_i^0 = \{0, 125, 123, 121, 117, 109\}$ $(i = 0, 1, \ldots, 5)$.

Store this information in matrix M:

$$M^T = \begin{pmatrix} 0 & 125 & 123 & 121 & 117 & 109 \\ 1 & 0 & 1 & 0 & 0 & 0 \end{pmatrix}.$$

We compute new positions and new bits to update the matrix M. The new bits are computed applying rule 102 to the terms of $\boldsymbol{s}$. They are $\{1 + 0, 0 + 1, 1 + 0, 0 +$

$0, 0 + 0\} = \{1, 1, 1, 0, \mathbf{0}\}$ to be stored, respectively, in the positions:

$$d_0^1 = \mathscr{L}_\alpha(\quad 0 - 125) + 125 = \mathscr{L}_\alpha(2) + 125 = 65$$
$$d_1^1 = \mathscr{L}_\alpha(125 - 123) + 123 = \mathscr{L}_\alpha(2) + 123 = 63$$
$$d_2^1 = \mathscr{L}_\alpha(123 - 121) + 121 = \mathscr{L}_\alpha(2) + 121 = 61$$
$$d_3^1 = \mathscr{L}_\alpha(121 - 117) + 117 = \mathscr{L}_\alpha(4) + 117 = 124$$
$$d_4^1 = \mathscr{L}_\alpha(117 - 109) + 109 = \mathscr{L}_\alpha(8) + 109 = 123.$$

Position 123 appears again and we have to store the value 0, but there is yet another bit in the same position with value 1. There is a discrepancy, so the guessed initial state $\boldsymbol{a}$ is not correct.

Output: $Stop = 0$. The initial state $\boldsymbol{a} = \{1, 1, 1, 1, 0, 1\}$ is not correct since a discrepancy has been found. ∎

In this example we had just to compute $\mathscr{L}_\alpha(1)$ since, due to the properties of Zech logarithm, $\mathscr{L}_\alpha(2) = 2 \cdot \mathscr{L}_\alpha(1)$, $\mathscr{L}_\alpha(4) = 4 \cdot \mathscr{L}_\alpha(1)$, $\mathscr{L}_\alpha(8) = 8 \cdot \mathscr{L}_\alpha(1)$. The computation of Zech logarithms is the most time-consuming part of the algorithm. Therefore, a certain number of logarithms, e.g., $L_1 - 1$ corresponding to the zero-run lengths in the PN-sequence generated by R_1, can be previously computed and stored in a table to be reused. At any rate, the properties of this discrete logarithm can be efficiently used to reduce the number of logarithms to be calculated [11, 47].

This algorithm requires an exhaustive search over 2^{L_1-1} initial states of R_1. In Table 4.6 some numerical results are depicted. We denote by $p_1(x)$ and $p_2(x)$ the characteristic polynomials of R_1 and R_2, respectively, n is the number of intercepted

Table 4.6 Some numerical results for the algorithm

$p_1(x)$	$p_2(x)$	n	T	N_{IS}
$1 + x^2 + x^3$	$1 + x^3 + x^4$	8	60	1
$1 + x^2 + x^3$	$1 + x^3 + x^5$	9	124	1
$1 + x^2 + x^5$	$1 + x + x^6$	11	1008	1
$1 + x^3 + x^5$	$1 + x + x^7$	13	2032	1
$1 + x^2 + x^5$	$1 + x^3 + x^7$	14	2032	1
$1 + x + x^6$	$1 + x^3 + x^7$	16	4046	1
$1 + x + x^7$	$1 + x^2 + x^3 + x^4 + x^8$	16	16,320	1
$1 + x + x^7$	$1 + x^4 + x^9$	16	32,704	1
$1 + x^2 + x^3 + x^4 + x^8$	$1 + x^4 + x^9$	17	65,408	1
$1 + x^4 + x^9$	$1 + x^3 + x^{10}$	18	261,888	1
$1 + x^4 + x^9$	$1 + x^2 + x^5 + x^9 + x^{10}$	19	261,888	1
$1 + x^2 + x^{11}$	$1 + x + x^5 + x^8 + x^{12}$	27	4,193,280	3
$1 + x^9 + x^{10} + x^{12} + x^{13}$	$1 + x + x^2 + x^5 + x^6 + x^{13} + x^{14}$	30	67,104,768	3
$1 + x^9 + x^{10} + x^{12}x^{13}$	$1 + x + x^4 + x^{15} + x^{16}$	52	268,431,360	1
$1 + x + x^2 + x^5 + x^6 + x^{13} + x^{14}$	$1 + x^2 + x^5 + x^{14} + x^{15}$	40	268,427,264	126
$1 + x^2 + x^5 + x^{14} + x^{15}$	$1 + x + x^4 + x^6 + x^{16}$	50	1,073,725,440	29
$1 + x + x^4 + x^{15} + x^{16}$	$1 + x + x^2 + x^6 + x^{10} + x^{11} + x^{17}$	58	4,294,934,528	206

bits, T represents the period of the corresponding shrunken sequence and N_{IS} denotes the number of R_1 initial states with no discrepancy. From these results, we can deduce that our algorithm presents two main advantages against other proposals.

The algorithm here presented is deterministic. It means that depending on the number of intercepted bits, the set of the possible correct states can have different sizes, but the correct one is certainly contained in such a set. This is an advantage against other probabilistic attacks. Besides, the required keystream length grows linearly in the length of R_2, while the period of the shrunken sequence grows exponentially. It means that the number of intercepted bits n needed for the attack is very low compared with the period of the shrunken sequence, fact that did not happen in other proposals like [50]. Low requirement of intercepted bits is a quite realistic condition for practical cryptanalysis.

The number of initial states for R_1 with no discrepancy is much smaller than the number of initial states analysed, so it simplifies the checking of the true pair of initial states in both registers R_1 and R_2.

Finally, this algorithm is particularly adequate for parallelization, that is, we can divide the 2^{L_1-1} possible initial states into different groups and process each group separately.

4.4   Different Attacks Against the Shrinking Generator: A Comparison

In Table 4.7, a comparison among distinct attacks against the shrinking generator is provided. In order to asses deterministic and probabilistic proposals, a typical correlation attack [90] is also included. This probabilistic attack requires computation of the joint probability for all possible R_2 initial states even though R_2 is the longest register. Moreover, to recover the R_1 initial state a further search (faster than 2^{L_1} but exponential in L_1) is also required.

In brief, according to Table 4.7 just the strategic bit attack exhibits jointly a linear computational complexity as well as a reduced number of intercepted bits, although these bits must be located at very specific positions.

In the sequel, different attacks against the self-shrinking generator will be the kernel of the following sections.

Table 4.7 Performance of different attacks against the shrinking generator

Attack	No. of intercepted bits	Computational complexity	Type of attack
Strategic bits	$N = L_2 \cdot L_1$	$O(L_2)$	Deterministic
LCT attack	$N = n \cdot L_2 \ (n = 2, 3, 4)$	$O(2^{L_1-1})$	Deterministic
Overlapping attack	$N = 2^{L_1-1}(2^{L_2} - \mathcal{Z}_\alpha(1))$	$O(L_2^3)$	Deterministic
102-CA recovery attack	$N = n \cdot L_2 \ (n = 2, 3, 4)$	$O(2^{L_1-1})$	Deterministic
Correlation attack	$N = 20 \cdot L_2$	$O(2^{L_2})$	Probabilistic

4.5 A Probabilistic Attack Against the Self-Shrinking Generator

As a representative example of probabilistic cryptanalysis against irregularly decimated generators, an algorithm for cryptanalysis of the self-shrinking generator [71]
is presented and discussed. In this paper, the self-shrinking generator is considered
under the following assumptions:

- The key of the cryptosystem is the initial state of the LFSR.
- The LFSR characteristic polynomial is known and there is no constraint on the
  number of LFSR feedback taps.
- The number of intercepted bits from the self-shrunken sequence is under a given
  limit (e.g., no more than 10^{10} bits).

According to these assumptions, this probabilistic approach is capable of finding the
LFSR initial state with probability close to 1. The work here considered improves,
in what computational complexity and constraints on the LFSR are concerned,
previous cryptanalysis found in the literature such as [36, 72] as well as it makes
use of the own self-decimation mechanism of the self-shrinking generator.

This attack, which can be classified as a "guess-and-determine" attack, uses the
intercepted self-shrunken sequence not only for hypothesis testing but also to reduce
the set of hypothesis to be tested.

The parameters of this approach are defined as:

- L is the LFSR length, for simplicity L is an even number.
- $\{a_i\}$, $i = 0, 1, 2, \ldots$, is the PN-sequence generated by the LFSR.
- $\{s_j\}$ ($0 \le j \le N - 1$) is the portion of intercepted self-shrunken sequence as
  well as N is the number of intercepted bits.
- $l < L/2$ is the supposed number of 1s in the current state, that is, a candidate for
  LFSR initial state.

In addition, $\{s_j\}$ is divided into $\lfloor (N - L)/l \rfloor$ l-dimensional successive but non-
overlapping segments $\mathbf{S}_k = [s_{kl+n}]$, ($k = 0, 1, \ldots, \lfloor (N - L)/l \rfloor - 1$), ($n =
0, 1, \ldots, l-1$). Recall that for an arbitrary pair of bits of the PN-sequence $\{a_i, a_{i+1}\}$,
if $a_i = 1$, then $a_{i+1} = s_n$, that is, a_{i+1} equals a term of the intercepted sequence;
otherwise, if $a_i = 0$, then $a_{i+1} = X$, that is, a_{i+1} equals an unknown bit of the
PN-sequence.

The basic idea of this attack is to test the possible LFSR states able to generate
a segment $\mathbf{S}_k$ of length l. Notice that, according to the generation rule of the self-
shrinking generator, these candidate states will have l 1s located at even positions
in the considered LFSR state. Roughly speaking, this approach consists in the
repetition of the following steps until the true initial state is reached:

1. Consider candidate states with l 1s at even positions and 0s at the remaining
   $(L/2 - l)$ even positions. The number of possible candidates with these characteristics is $\binom{L/2}{l}$.

2. For each l-dimensional segment $\mathbf{S}_k$ $(k = 0, 1, \ldots, \lfloor(N - L)/l\rfloor - 1)$ of the self-shrunken sequence do the following:

   - Put the l values s_n of $\mathbf{S}_k$ at the odd positions next to the positions with 1. Thus, for each candidate $2l + (L/2 - l)$ state positions are already determined (l positions filled with 1s, l positions with values s_n and $(L/2 - l)$ positions with $0's$).
   - Give binary values to the remaining $(L/2 - l)$ not filled positions, in total $2^{(L/2 - l)}$ different binary configurations.
   - Check if among the constructed candidate states is the true one.

The appropriate choice of the parameter l in step 1 could yield probabilistic reduction of the number of hypothesis to be tested in step 2.

In [71], the recommended value for the parameter l is the maximum integer l such that

$$1 \leq \lfloor(N - L)/l\rfloor 2^{-L/2}\binom{L/2}{l}.$$

According to the paper [71], this approach ensures the cryptanalysis with overall complexity 2^{L-l} assuming that the required length N of intercepted sequence is upper bounded by

$$N \leq l\, 2^{L/2}\binom{L/2}{l}.$$

In a more detailed way, the previous algorithm can be described as follows:

Input: The LFSR characteristic polynomial and the N bits of intercepted sequence $\{s_j\}$ $(0 \leq j \leq N - 1)$.

Initialization: Determine the value of the parameter l and divide the sequence $\{s_j\}$ into $\lfloor(N - L)/l\rfloor$ l-dimensional successive non-overlapping segments $\mathbf{S}_k$.

For each one of the $\binom{L/2}{l}$ different locations of l 1s do steps 1–5.

Step 1: Set l 1s at even positions and 0s at the remaining $(L/2 - l)$ even positions of the current state.

For each segment $\mathbf{S}_k$ $(k = 0, 1, \ldots, \lfloor(N - L)/l\rfloor - 1)$, do steps 2–5.

Step 2: Put the value s_{kl+n} at the odd position whose left-side neighbour position contains the $(n + 1)$th 1 $(n = 0, 1, \ldots, l - 1)$ in the segment $\mathbf{S}_k$.

For each one of the $2^{(L/2-l)}$ different binary configurations, do steps 3–5.

Step 3: Put a particular binary configuration at the odd positions not filled in step 2.

Step 4: According to the current LFSR initial content determined in steps 1–3, generate $L^* \geq L$ output bits from the self-shrinking generator.

Step 5: If the L^*-segment generated in step 4 is equal to the segment $[s_{(k+1)l+n}]$ $(n = 0, 1, \ldots, L^* - 1)$, then conclude that the initial state (secret key) is reconstructed and Stop. Otherwise continue the searching.

Output: The reconstructed initial state (secret key) or the conclusion that the solution is not found because of the probabilistic nature of the algorithm.

The previous cryptanalysis is a probabilistic attack. According to [71], it can be stated that:

1. The probability p that the set of hypothesis tested by the algorithm contains the true one is

$$p = 1 - (1 - 2^{-L/2} \binom{L/2}{l})^{\lfloor (N-L)/l \rfloor}.$$

So the proposed algorithm allows with probability close to 1 the reconstruction of the secret key assuming a sufficiently long intercepted segment $\{s_j\}$ ($0 \le j \le N - 1$).

2. The number of hypothesis H to be tested is upper bounded by

$$H \le \binom{L/2}{l} \lfloor (N-L)/l \rfloor 2^{L/2-l}.$$

3. The expected number of hypothesis $\bar{H}$ to be tested is upper bounded by

$$\bar{H} \le 2^{L-l}.$$

4. The expected gain g obtained by this algorithm compared with the attack proposed in [67] is lower bounded by

$$g > \frac{2^{0.75L}}{2^{L-l}} = 2^{l-.25L}.$$

The complexity of the above algorithm is determined by the number of hypothesis to be tested. In Table 4.8 the upper bound on the algorithm complexity C_{ub} and the expected lower bound on the gain g_{lb} are depicted for different LFSR length L and intercepted sequence length N. The algorithm performance does not depend

Table 4.8 Complexity and gain of this algorithm for different values of L and N

L	N	C_{ub}	g_{lb}
60	1.88×10^5	2^{35}	2^{10}
60	6.96×10^7	2^{32}	2^{13}
80	5.85×10^7	2^{45}	2^{15}
80	4.11×10^9	2^{43}	2^{17}
100	4.86×10^8	2^{57}	2^{18}
100	3.12×10^9	2^{56}	2^{19}

on the number of LFSR feedback taps and makes use of a length of intercepted sequence under a given limit. The attack succeeds if and only if any of the internal state guess matches the corresponding l internal state bits that generated the segment of intercepted bits under consideration.

4.5.1 Other Probabilistic Attacks Against the Self-Shrinking Generator

There are other cryptanalytic attacks performed over this generator which need a different amount of intercepted bits. In particular, two other guess-and-determine attacks are next discussed. The basic idea in these examples is to guess some bits of the LFSR internal state and derive the remaining bits from the intercepted keystream sequence. Then, the acceptance or rejection of each guessed internal state must be individually checked.

- In [105], a new approach is developed. Now the algorithm requires less than $2^{L/2}$ intercepted bits, the knowledge of the LFSR characteristic polynomial and a block of $2^{0.25 \cdot L}$ LFSR internal state guesses with binary values assigned to l bits ($l < L/2$). For each one of these guesses do:

  (a) Write out a system of linear equations in the remaining $L - l$ LFSR internal state bits by using the intercepted bits;
  (b) Check the consistency of such an equation system; and
  (c) If the linear consistency is confirmed, then solve the linear system and generate keystream sequence from this internal state. If the generated sequence matches the intercepted bits, then the guess is accepted as LFSR internal state. Otherwise shift the keystream by one bit forward and repeat the process. If after the successive shifts along the intercepted keystream sequence there is no matching solution, then try a new LFSR internal state guess and go again to step (a).

- In [81], a cryptanalytic attack based on the previous one is described. Compared with [105], reference [81] is a more refined procedure in what the choice of internal state guesses is concerned. Table 4.9 shows a comparison of parameters for the different authors when the LFSR length is $L = 40$. In addition, Q is the number of internal state guesses and N the number of intercepted bits. Moreover, Table 4.10 introduces a comparison of complexities (time, memory and data) for the different cryptanalysis.

Table 4.9 Parameter comparison among different authors for $L = 40$

Author	Q	N
Mihaljevic ($l = 20$)	$2^{L-l} = 1,048,576$	10^6
Pazo-Robles et al. ($l = 20$)	2736	$700 - 800$
Zhang et al. ($l = 25$)	2^{22}	$O(2^8)$

Author	C_T	C_M	C_D
Mihaljevic	$O(2^{0.7*L})$	$O(L)$	$O(2^{0.5*L})$
Pazo-Robles et al.	$O(2^{0.6*L})$	$O(L^2)$	$< O(2^{0.25*L})$
Zhang et al.	$O(2^L)$	$O(L^2)$	$O(2^{0.2*L})$

Table 4.10 Complexities for the different authors

4.6 CA-Based Linearization Attack Against the Self-Shrinking Generator

In Chap. 3, it was seen that the output sequences obtained from decimation-based generators can be modelled by means of linear 102-CA (60-CA). Although this linearization procedure can be applied to different sequence generators (see Chap. 3), in this section we focus exclusively on the self-shrinking generator. Indeed, in [8, Theorem 2] it is proved that the self-shrunken sequence $\{s_j\}$ $j = 0, 1, 2, \ldots$, is generated by an uniform null linear 102-CA (60-CA) of length n. The idea of this section is to exploit the linearity of these cellular structures in order to recover the self-shrunken sequence from a certain amount of intercepted bits.

In this attack, the self-shrinking generator is considered under the following assumptions:

- The key of the cryptosystem is the initial state of the LFSR.
- The LFSR characteristic polynomial is not needed.
- The number of intercepted bits from the self-shrunken sequence is the length n of the CA.

According to these assumptions, this deterministic approach determines the CA initial state just performing simple XOR operations. Once such an initial state is known, the rules 102/60 allow one the generation of the whole self-shrunken sequence. It is worth noticing that in this cellular scenario, the output sequence is recovered without knowledge of the cryptosystem key. Let us see an illustrative example of generation of the self-shrunken sequence by means of rules 102 and 60.

Example 4.2 Given an LFSR with length $L = 4$, characteristic polynomial $p(x) = 1 + x^3 + x^4 \in \mathbb{F}_2[x]$ and initial state $\{1, 0, 1, 0\}$, the self-shrunken sequence obtained is $\{s_j\} = \{00101101\}$, with period $T = 2^3$. The characteristic polynomial of this sequence is $p_5(x) = (1 + x)^5$. In Table 4.11, a double example of one-dimensional linear CAs of length 5 is depicted. Table 4.11a represents an uniform null linear 102-CA where the sequence in bold at the first column (numbered from left to right) is the self-shrunken sequence. It is easy to check that the characteristic polynomials of the remaining sequences are $p_t(x) = (x + 1)^t$ ($t = 4, 3, 2, 1$), respectively.

Table 4.11 CAs that generate the self-shrunken sequence in Example 4.2

(a) Uniform null linear 102-CA					(b) Uniform null linear 60-CA				
102	102	102	102	102	60	60	60	60	60
0	0	1	1	1	1	1	1	0	0
0	1	0	0	1	1	0	0	1	0
1	1	0	1	1	1	1	0	1	1
0	1	1	0	1	1	0	1	1	0
1	0	1	1	1	1	1	1	0	1
1	1	0	0	1	1	0	0	1	1
0	1	0	1	1	1	1	0	1	0
1	1	1	0	1	1	0	1	1	1

The same CA of length 5 with rule 60 is depicted in Table 4.11b. Now the output sequences will be the same as before but in reverse order. ∎

In brief, a linear 102-CA (60-CA) starting at a particular initial state models the previous self-shrunken sequence.

In the following subsections, distinct features of this cryptanalysis are considered. In fact, two different methods of determining the CA initial state have been developed. Next, the number of XOR operations needed to recover the whole self-shrunken sequence is computed. The last subsection exhibits a comparison between the CAs used in this work and those ones proposed in [33].

4.6.1 Computation of the CA Initial State

In order to simplify the notation, it is assumed that the linear complexity of the self-shrunken sequence is denoted by n. From the previous section, we know that there exists a linear 102-CA of length n that generates the self-shrunken sequence and whose last sequence (numbered from left to right) is the identically 1 sequence. Assume that $n - 1$ bits of the self-shrunken sequence $\{s_0, s_1, \ldots, s_{n-2}\}$ are intercepted. These bits correspond to a portion of the sequence generated by the CA at the first column. Two different methods of computing the CA initial state are now described:

Method 1 From the $(n - 1)$ intercepted bits and according to rule 102, we can compute $(n - 2)$ bits of the sequence located at the 2nd column that will be $\{s_0 + s_1, s_1 + s_2, \ldots, s_{n-3} + s_{n-2}\}$ performing $(n - 2)$ XOR operations. Next, we can compute $(n - 3)$ bits of the sequence located at the 3rd column, $\{s_0 + s_2, s_1 + s_3, \ldots, s_{n-4} + s_{n-2}\}$ performing $(n - 3)$ XOR operations and so on. Proceeding in this way, the CA initial state $\{s_0, s_0 + s_1, s_0 + s_2, \ldots, 1\}$ (the first bit of each vertical sequence) is obtained just making use of XOR operations. Therefore, it is enough to know $(n - 1)$ bits of the self-shrunken sequence (the first vertical sequence) to compute the CA initial state, and then from it to recover the remaining bits.

Table 4.12 Necessary bits to recover the CA initial state

102	102	102	102	102
0	0	1	1	1
0	1	0		1
1	1			1
0				1
				1
				1
				1
				1

Furthermore, the number of XOR operations needed to compute the CA initial state is

$$\sum_{i=1}^{n-2} i = \frac{(n-1)(n-2)}{2}.$$

In Example 4.2, the self-shrunken sequence had period $T = 8$ and linear complexity $LC = 5$. In Table 4.12, we can see that given 4 intercepted bits of the self-shrunken sequence (the bits in bold at the first column), we can reconstruct portions of the other sequences to compute the CA initial state $\{0, 0, 1, 1, 1\}$. As $n = 5$, we need to perform six XOR operations to recover the CA initial state.

Method 2 In this case, a general expression for each bit of the CA initial state is deduced. Assume $\{s_i\}$ is the sequence at the first column of the CA. Let N be a positive integer and let $i_0, i_1, i_2, \ldots, i_t$ be binary coefficients for the binary representation of N, that is, $N = i_0 2^0 + i_1 2^1 + i_2 2^2 + \cdots + i_t 2^t$. Define $J = \{k \mid i_k \neq 0, k = 0, 1, 2, \ldots t\}$, with $|J| = m$ and $m \leq t$. Now denote the elements of $J = \{j_0, j_1, \ldots, j_{m-1}\}$, then,

$$N = 2^{j_0} + 2^{j_1} + \cdots + 2^{j_{m-1}}.$$

The first bit of the $(N + 1)$th sequence (the $(N + 1)$th bit of the CA initial state) is given by

$$
s_0 + \sum_{l=0}^{m-1} s_{2^{j_l}} + \sum_{l_1=0}^{m-1} \sum_{\substack{l_2=0 \\ l_2 \neq l_1}}^{m-1} s_{2^{j_{l_1}} + 2^{j_{l_2}}}
$$

$$
+ \sum_{l_1=0}^{m-1} \sum_{\substack{l_2=0 \\ l_2 \neq l_1}}^{m-1} \sum_{\substack{l_3=0 \\ l_3 \neq l_1 \\ l_3 \neq l_2}}^{m-1} s_{2^{j_{l_1}} + 2^{j_{l_2}} + 2^{j_{l_3}}} + \cdots + s_N. \tag{4.7}
$$

In the same way, the second bit of the $(N+1)$th sequence is given by

$$s_1 + \sum_{l=0}^{m-1} s_{2^{jl}+1} + \sum_{l_1=0}^{m-1} \sum_{\substack{l_2=0 \\ l_2 \neq l_1}}^{m-1} s_{2^{jl_1}+2^{jl_2}+1}$$

$$+ \sum_{l_1=0}^{m-1} \sum_{\substack{l_2=0 \\ l_2 \neq l_1}}^{m-1} \sum_{\substack{l_3=0 \\ l_3 \neq l_1 \\ l_3 \neq l_2}}^{m-1} s_{2^{jl_1}+2^{jl_2}+2^{jl_3}+1} + \cdots + s_{N+1}. \tag{4.8}$$

According to the rule 102, the first bit of the $(N+2)$th sequence (the $(N+2)$th bit of the CA initial state) is given by Eq. (4.7) plus Eq. (4.8).

As a simple numerical example, let $N = 12$. Since $N = 4 + 8$, the 13th bit will have the form $s_0 + s_4 + s_8 + s_{12}$. Furthermore, the 14th bit will have the form $s_0 + s_4 + s_8 + s_{12} + s_1 + s_5 + s_9 + s_{13}$. In [8, Appendix A], two arrays with the generation of the first nine columns of the CA and the generation of the first nine bits of the CA initial state are depicted. Both expressions can be proven by induction over the number N, see [8].

4.6.2 Reconstruction of the Self-Shrunken Sequence: The Number of XOR Operations

Given the CA initial state of length n and considering that the period of the self-shrunken sequence is 2^{L-1}, we would like to compute the number of XOR operations needed to reconstruct the self-shrunken sequence. We know that the last sequence is the identically 1 sequence. Next sequence (from right to left) is uniquely determined by the first bit, that is, if this bit is 0, then the sequence is $\{010101\ldots\}$. Otherwise, if this bit is 1, then the sequence is $\{101010\ldots\}$.

According to [8, Theorem 4], the next two sequences have period 4. Since the first bit of both sequences (bits of the CA initial state) is known and according to rule 102, just three XOR operations must be performed to compute the remaining three bits (one operation per bit). That is, we have to perform six operations to thoroughly compute both sequences.

In general, according to [8, Theorem 4], there are: 2^{i-1} sequences with period 2^i for $(i = 2, \ldots, L-2)$ and $n - 2^{L-2}$ sequences with period 2^{L-1}. Therefore, following the same procedure as before, it is clear that the number of XOR operations needed to recover the first 2^{L-1} bits of the self-shrunken sequence is given by

$$\sum_{i=2}^{L-2} (2^i - 1)2^{i-1} + (n - 2^{L-2})(2^{L-1} - 1) = \frac{3n(2^{L-1} - 1) - 2^{2L-2} - 2}{3}. \tag{4.9}$$

Fig. 4.1 Necessary XOR
operations needed to compute
the self-shrunken sequence

In Table 4.11a, it is showed that the computed CA initial state is $\{0, 0, 1, 1, 1\}$. In this case, we know that $n = 5$ and $L = 4$ (from Example 4.2). Then, according to expression (4.9), we have to perform 13 XOR operations. These operations are depicted in Fig. 4.1. The bits on grey rectangles are the portions of sequences we can deduce without computing any operation.

4.6.3 102/60 vs. 90/150 CA Proposals

In [33], the authors propose a family of CAs based on rules 90/150 that generate the self-shrunken sequence too. Such CAs have a well-defined structure; rule 90 is applied to the extreme cells, while rule 150 to the remaining cells. In the present work, the CAs have a very specific structure as well. Indeed, the last column is the identically 1 sequence. Besides, there is always a sequence of period 2 (the sequence $\{0101\ldots\}$ or the sequence $\{1010\ldots\}$). After this, there are 2 sequences of period 4, 2^2 sequences of period 8 and so on, until we get 2^{L-3} sequences with period 2^{L-2}, L being the LFSR length. The remaining sequences (the length of the CA minus 2^{L-2}) have period 2^{L-1}, including the self-shrunken sequence. On the other hand, we know that the linear complexity n of the self-shrunken sequence satisfies the inequality $2^{L-2} < n \leq 2^{L-1} - (L - 2)$. Hence the length of these CAs is less than 2^{L-1}, the length of those ones proposed in [33]. For instance, in order to model the self-shrunken sequence in Example 4.2, we need a CA of length 5 (see Table 4.11a). In case of using the CA proposed in [33], the CA length would have been 8. As far as L increases the length of the 102/60 CAs is smaller than that of the 90/150 CAs. Furthermore, taking into account that the rule 102 is based on the XOR of 2 bits, while the rule 150 is based on the XOR of 3 bits [33], the amount of operations required to recover the self-shrunken sequence with the CAs here proposed is significantly smaller.

Currently the linear complexity of a sequence is computed by means of the Berlekamp–Massey algorithm. In order to compute the linear complexity n of a sequence, this algorithm needs to analyse $2n$ bits of such a sequence. Hence, recall that the amount of intercepted bits $(n - 1)$ needed by the 102/60 CAs to recover the whole sequence is half the bits needed by the Berlekamp–Massey algorithm.

4.7 Different Attacks Against the Self-Shrinking Generator: A Comparison

The class of guess-and-determine attacks against the self-shrinking generator is characterized by:

- They are probabilistic attacks.
- They need about $2^{L/2}$ intercepted bits.
- The knowledge of the LFSR characteristic polynomial (which is recommended to be a part of the key) is needed.
- They involve a large computational complexity to check the number of guessed initial states.

On the other hand, the method of sequence reconstruction based on 102/60 CAs is characterized by:

- It is a deterministic attack.
- It requires a much greater amount of intercepted bits ($2^{L-2} < n < 2^{L-1}$) than the previous attacks.
- The knowledge of the LFSR characteristic polynomial (which is recommended to be a part of the key) is not needed.
- The computation is performed exclusively by means of XOR logic operations.

Thus, both types of cryptanalytic attacks are suitable only against self-shrinking generators where L is not large. Consequently, the self-shrinking generator cannot be considered practically secure if the underlying LFSR length is small.

References

1. Babbage, S., Dodd, M.: The MICKEY stream ciphers. In: Robshaw, M., Billet, O. (eds.) New Stream Cipher Designs. The eSTREAM Finalists. Lecture Notes in Computer Science, vol. 4986, pp. 191–209. Springer, Berlin, Heidelberg (2008). https://doi.org/10.1007/978-3-540-68351-3_15
2. Bertoni, G., Daemen, J., Peeters, M., Van Assche, G.: Cryptographic sponge functions (2011). https://keccak.team/sponge_duplex.html
3. Beth, T., Piper, F.C.: The Stop-and-go generator. In: Beth, T., Cot, N., Ingemarsson, I. (eds.) Advances in Cryptology, Proceedings of EUROCRYPT'84. Lecture Notes in Computer Science, vol. 209, pp. 88–92. Springer, Berlin, Heidelberg (1985). https://doi.org/10.1007/3-540-39757-4_9
4. Biryukov, A., Shamir, A., Wagner, D.: Real time cryptanalysis of A5/1 on a PC. In: Goos, G., Hartmanis, J., Van Leeuwen, Schneier, B. (eds.) Proceedings of Fast Software Encryption 2000. Lecture Notes in Computer Science, vol. 1978, pp. 1–18. Springer, Berlin, Heidelberg (2001). https://doi.org/10.1007/3-540-44706-7_1
5. Blackburn, S.R.: The linear complexity of the self-shrinking generator. IEEE Trans. Inf. Theory **45**(6), 2073–2077 (1999). https://doi.org/10.1109/18.782139
6. Briceno, M., Goldberg, I., Wagner, D.: A pedagogical implementation of A5/1 (1999). http://www.scard.org/gsm/a51.html
7. Cardell, S.D., Climent, J.J.: A construction of primitive polynomials over finite fields. Linear Multilinear Algebra **65**(12), 2424–2431 (2017). https://doi.org/10.1080/03081087.2016.1275507
8. Cardell, S.D., Fúster-Sabater, A.: Linear models for the self-shrinking generator based on CA. J. Cell. Automata **11**(2–3), 195–211 (2016)
9. Cardell, S.D., Fúster-Sabater, A.: Modelling the shrinking generator in terms of linear CA. Adv. Math. Commun. **10**(4), 797–809 (2016). https://doi.org/10.3934/amc.2016041
10. Cardell, S.D., Fúster-Sabater, A.: Discrete linear models for the generalized self-shrunken sequences. Finite Fields Appl. **47**, 222–241 (2017). https://doi.org/10.1016/j.ffa.2017.06.010
11. Cardell, S.D., Fúster-Sabater, A.: Recovering decimation-based cryptographic sequences by means of linear CAs (2018). https://arxiv.org/abs/1802.02206
12. Cardell, S.D., Fúster-Sabater, A., Ranea, A.: Linearity in decimation-based generators: an improved cryptanalysis on the shrinking generator. Open Math. **16**(1), 646–655 (2018). https://doi.org/10.1515/math-2018-0058
13. Cattell, K., Muzio, J.C.: One-dimensional linear hybrid cellular automata. IEEE Trans. Comput.-Aided Des. **15**(3), 325–335 (1996). https://doi.org/10.1109/43.489103

14. Cid, C., Kiyomoto, S., Kurihara, J.: The RAKAPOSHI stream cipher. In: Qing, S., Mitchell, C., Wang, G. (eds.) Information and Communications Security. ICICS 2009. Lecture Notes in Computer Science, vol. 5927, pp. 32–46. Springer, Berlin, Heidelberg (2009). https://doi.org/10.1007/978-3-642-11145-7_5

15. Coppersmith, D., Krawczyk, H., Mansour, Y.: The shrinking generator. In: Stinson, D. (ed.) Advances in Cryptology – CRYPTO '93. Lecture Notes in Computer Science, vol. 773, pp. 22–39. Springer, Berlin (1994). https://doi.org/10.1007/3-540-48329-2_3

16. Courtois, N.T.: Fast algebraic attacks on stream ciphers with linear feedback. In: Boneh, D. (ed.) Advances in Cryptology – CRYPTO 2003. Lecture Notes in Computer Science, vol. 2729, pp. 176–194. Springer, Berlin, Heidelberg (2003)

17. Courtois, N.T., Meier, W.: Algebraic attacks on stream ciphers with linear feedback. In: Biham, E. (ed.) Advances in Cryptology – EUROCRYPT 2003. Lecture Notes in Computer Science, vol. 2656, pp. 345–359. Springer, Berlin (2003)

18. Courtois, N.T., O'Neil, S., Quisquater, J.J.: Practical algebraic attacks on the Hitag2 stream cipher. In: Samarati, P., Yung, M., Martinelli, F., Ardagna, C.A. (eds.) Information Security – ISC 2009. Lecture Notes in Computer Science, vol. 5735, pp. 167–176. Springer, Berlin, Heidelberg (2009). https://doi.org/10.1007/978-3-642-04474-8_14

19. Daemen, J., Rijmen, V.: The Design of Rinjdael. Springer, Berlin (2002)

20. Das, S., RoyChowdhury, D.: Car30: a new scalable stream cipher with rule 30. Cryptography and Communications **5**(2), 137–162 (2013). https://doi.org/10.1007/s12095-012-0079-1

21. Diffie, W.D., Hellman, M.E.: New directions in cryptography. IEEE Trans. Inf. Theory **22**(6), 644–654 (1976). https://doi.org/10.1109/TIT.1976.1055638

22. Duvall, P.F., Mortick, J.C.: Decimation of periodic sequences. SIAM J. Appl. Math. **21**(3), 367–372 (1971). https://doi.org/10.1137/0121039

23. Ekdahl, P., Meier, W., Johansson, T.: Predicting the shrinking generator with fixed connections. In: Biham, E. (ed.) Advances in Cryptology-EUROCRYPT'2003. Lecture Notes in Computer Science, vol. 2656, pp. 330–344. Springer, Berlin, Heidelberg (2003). https://doi.org/10.1007/3-540-39200-9_20

24. Fluhrer, S., Lucks, S.: Analysis of the E_0 encryption system. In: Vaudenay, S., Youssef, A. (eds.) Selected Areas in Cryptography SAC 2001. Lecture Notes in Computer Science, vol. 2259, pp. 38–48. Springer, Berlin, Heidelberg (2001). https://doi.org/10.1007/3-540-45537-X_3

25. Fúster-Sabater, A.: Run distribution in nonlinear binary generators. Appl. Math. Lett. **17**(12), 1427–1432 (2005). https://doi.org/10.1016/j.aml.2002.09.003

26. Fúster-Sabater, A.: Computing classes of cryptographic sequence generators. In: Proceedings of International Conference on Computational Science (ICCS2013). Procedia Computer Science, vol. 18, pp. 2440–2443 (2013). https://doi.org/10.1016/j.procs.2013.05.419

27. Fúster-Sabater, A., Caballero-Gil, P.: On the linear complexity of nonlinearly filtered PN-sequences. In: Pieprzyk, J., Safavi-Naini, R. (eds.) Advances in Cryptology, ASIACRYPT'94. Lecture Notes in Computer Science, vol. 917, pp. 80–90. Springer, Berlin (1995). https://doi.org/10.1007/BFb0000426

28. Fúster-Sabater, A., Caballero-Gil, P.: Linear solutions for cryptographic nonlinear sequence generators. Phys. Lett. A **369**(5–6), 432–437 (2007). https://doi.org/10.1016/j.physleta.2007.04.103

29. Fúster-Sabater, A., Caballero-Gil, P.: Strategic attack on the shrinking generator. Theor. Comput. Sci. **409**(3), 530–536 (2008). https://doi.org/10.1016/j.tcs.2008.09.030

30. Fúster-Sabater, A., Caballero-Gil, P.: Chaotic modelling of the generalized self-shrinking generator. Appl. Soft Comput. **11**(2), 1876–1880 (2011). https://doi.org/10.1016/j.asoc.2010.06.002

31. Fúster-Sabater, A., García-Mochales, P.: On the balancedness of nonlinear generators of binary sequences. Inf. Process. Lett. **85**(2), 111–116 (2003). https://doi.org/10.1016/S0020-0190(02)00349-6

32. Fúster-Sabater, A., García-Mochales, P.: A simple computational model for acceptance/rejection of binary sequence generators. Appl. Math. Model. **31**(8), 1548–1558 (2007)

33. Fúster-Sabater, A., Pazo-Robles, M.E., Caballero-Gil, P.: A simple linearization of the self-shrinking generator by means of cellular automata. Neural Netw. **23**(3), 461–464 (2010). https://doi.org/10.1016/j.neunet.2009.12.008

34. Gayoso, V., Hernández, L., Martín, A., Zhang, J.: Breaking a Hitag2 protocol with low cost technology. In: Proceedings of the 3rd International Conference on Information Systems Security and Privacy, ICISSP 2017, Porto, pp. 579–584 (2017). https://doi.org/10.5220/0006271905790584

35. Golić, J.D.: Embedding and probabilistic correlation attacks on clock-controlled shift registers. In: De Santis, A. (ed.) Advances in Cryptology-EUROCRYPT'94. Lecture Notes in Computer Science, vol. 950, pp. 230–243. Springer, Berlin, Heidelberg (1995). https://doi.org/10.1007/BFb0053439

36. Golić, J.D.: Towards fast correlation attacks on irregularly clocked shift registers. In: Guillou, L., Quisquater, J. (eds.) Advances in Cryptology-EUROCRYPT '95. Lecture Notes in Computer Science, vol. 921, pp. 248–261. Springer, Berlin, Heidelberg (1995). https://doi.org/10.1007/3-540-49264-X_20

37. Golić, J.D.: Cryptanalysis of alleged A5 stream cipher. In: Fumy, W. (ed.) International Conference on the Theory and Application of Cryptographic Techniques, Advances in Cryptology – EUROCRYPT '97. Lecture Notes in Computer Science, vol. 1233, pp. 239–255. Springer, Berlin, Heidelberg (1997). https://doi.org/10.1007/3-540-69053-0_17

38. Golić, J.D.: Iterative optimum symbol-by-symbol decoding and fast correlation attacks. IEEE Trans. Inf. Theory **47**(7), 3040–3049 (2001). https://doi.org/10.1109/18.959285

39. Golić, J.D., Mihaljević, M.J.: A generalized correlation attack on a class of stream ciphers based on the Levenshtein distance. J. Cryptol. **3**(3), 201–212 (1991). https://doi.org/10.1007/BF00196912

40. Golić, J.D., Petrovic, S.V.: Correlation attacks on clock-controlled shift registers in keystream generators. IEEE Trans. Comput. **45**(4), 482–486 (1996). https://doi.org/10.1109/12.494106

41. Golomb, S.W.: Shift Register-Sequences. Aegean Park Press, Laguna Hill (1982)

42. Gomulkiewicz, M., Kutylowski, M., Wlaź, P.: Fault cryptanalysis and the shrinking generator. In: Álvarez, C., Serna, M. (eds.) 5th International Workshop on Experimental Algorithms (WEA 2006). Lecture Notes in Computer Science, vol. 4007, pp. 61–72. Springer, Berlin, Heidelberg (2006). https://doi.org/10.1007/11764298_6

43. Günther, C.: Alternating step generators controlled by the Bruijn sequences. In: Chaum, D., Price, W. (eds.) Advances in Cryptology, Proceedings of EUROCRYPT'87. Lecture Notes in Computer Science, vol. 304, pp. 5–14. Springer, Berlin, Heidelberg (1987). https://doi.org/10.1007/3-540-39118-5_2

44. Hell, M., Johansson, T.: Breaking the F-FCSR-H stream cipher in real time. In: Pieprzyk, J. (ed.) Advances in Cryptology – ASIACRYPT 2008. Lecture Notes in Computer Science, vol. 5350, pp. 557–569. Springer, Berlin, Heidelberg (2008). https://doi.org/10.1007/978-3-540-89255-7_34

45. Hernández, C., Fúster-Sabater, A.: Deterministic analysis of balancedness in symmetric cryptography. In: Levi, A., Savaş, E., Yenigün, H., Balcısoy, S., Saygın, Y. (eds.) Computer and Information Sciences – ISCIS 2006. Lecture Notes in Computer Science, vol. 4263, pp. 1011–1020. Springer, Berlin (2006). https://doi.org/10.1007/11902140_105

46. Hu, Y., Xiao, G.: Generalized self-shrinking generator. IEEE Trans. Inf. Theory **50**(4), 714–719 (2004). https://doi.org/10.1109/TIT.2004.825256

47. Huber, K.: Some comments on Zech's logarithms. IEEE Trans. Inf. Theory **36**(4), 946–950 (1990). https://doi.org/10.1109/18.53764

48. Jacobi, C.G.J.: Über die Kreisteilung und ihre Anwendung auf die Zahlentheorie. J. Reine Angew. Math. **30**, 166–182 (1846)

49. Jenkins, C., Schulte, M., Glossner, J.: Instructions and hardware designs for accelerating SNOW 3G on a software-defined radio platform. Analog Integr. Circuits Signal Process. **69**(2–3), 207–218 (2011). https://doi.org/10.1007/s10470-011-9712-8

50. Johansson, T.: Reduced complexity correlation attacks on two clock-controlled generators. In: Ohta, K., Pei, D. (eds.) Advances in Cryptology – ASIACRYPT'98. *Lecture Notes in Computer Science*, vol. 1514, pp. 342–357. Springer, Berlin, Heidelberg (1998)

51. Jose, J., Das, S., RoyChowdhury, D.: Inapplicability of fault attacks against trivium on a cellular automata based stream cipher. In: 11th International Conference on Cellular Automata for Research and Industry, ACRI 2014. Lecture Notes in Computer Science, vol. 8751, pp. 427–436. Springer, Cham (2014). https://doi.org/10.1007/978-3-319-11520-7_44

52. Kahn, D.: The Codebreakers. The Story of Secret Writing. Macmillan, New York (1967)

53. Kanso, A.: Modified self-shrinking generator. Comput. Electr. Eng. **36**(5), 993–1001 (2010). https://doi.org/10.1016/j.compeleceng.2010.02.004

54. Kiyomoto, S., Tanaka, T., Sakurai, K.: K2: a stream cipher algorithm using dynamic feedback control. In: Proceedings of the International Conference on Security and Cryptography – SECRYPT 2007, pp. 204–213 (2008)

55. Kolokotronis, N., Kalouptsidis, N.: On the linear complexity of nonlinearly filtered pn-sequences. IEEE Trans. Inf. Theory **49**(11), 3047–3059 (2003). https://doi.org/10.1109/TIT.2003.818400

56. Kolokotronis, N., Limniotis, K., Kalouptsidis, N.: Lower bounds on sequence complexity via generalised vandermonde determinants. In: Gong, G., Helleseth, T., Song, H.Y., Yang, K. (eds.) Sequences and Their Applications - SETA 2006. Lecture Notes in Computer Science, vol. 4086, pp. 271–284. Springer, Berlin, Heidelberg (2006). https://doi.org/10.1007/11863854_23

57. Krawczyk, H.: The shrinking generator: some practical considerations. In: Anderson, R. (ed.) International Workshop on Fast Software Encryption-FSE'93. Lecture Notes in Computer Science, vol. 809, pp. 45–46. Springer, Berlin, Heidelberg (1994). https://doi.org/10.1007/3-540-58108-1_5

58. Liberti, L.: Structure of the invertible CA transformations group. J. Comput. Syst. Sci. **59**(3), 521–536 (1999). https://doi.org/10.1006/jcss.1999.1659

59. Lidl, R., Niederreiter, H.: Introduction to Finite Fields and Their Applications. Cambridge University Press, New York (1986)

60. Limniotis, K., Kolokotronis, N., Kalouptsidis, N.: On the linear complexity of sequences obtained by state space generators. IEEE Trans. Inf. Theory **54**(4), 1786–1793 (2008). https://doi.org/10.1109/TIT.2008.917639

61. Marsaglia, G.: The Marsaglia Random Number CDROM including the Diehard battery of tests of randomness, Florida State University (1995). http://webhome.phy.duke.edu/~rgb/General/dieharder.php

62. Marsaglia, G., Tsang, W.: Some difficult-to-pass tests of randomness. J. Stat. Softw. **7**(3) (2002). https://doi.org/10.18637/jss.v007.i03. https://www.jstatsoft.org/article/view/v007i03

63. Massey, J.L.: Shift-register synthesis and BCH decoding. IEEE Trans. Inf. Theory **15**(1), 122–127 (1969). https://doi.org/10.1109/TIT.1969.1054260

64. Massey, J.L.: An introduction to contemporary cryptology. Proc. IEEE **76**(5), 533–549 (1988). https://doi.org/10.1109/5.4440

65. Massey, J.L., Rueppel, R.A.: Linear ciphers and random sequence generators with multiple clocks. In: Beth, T., Cot, N., Ingemarsson, I. (eds.) Advances in Cryptology – EURO-CRYPT'84. Lecture Notes in Computer Science, vol. 209, pp. 74–87. Springer, Berlin, Heidelberg (1985). https://doi.org/10.1007/3-540-39757-4_8

66. Meier, W., Staffelbach, O.: Analysis of pseudo random sequences generated by cellular automata. In: Davies, D. (ed.) Analysis of Pseudo Random Sequences Generated by Cellular Automata. Advances in Cryptology – EUROCRYPTO '91. Lecture Notes in Computer Science, vol. 547, pp. 186–199. Springer, Berlin, Heidelberg (1991). https://doi.org/10.1007/3-540-46416-6_17

67. Meier, W., Staffelbach, O.: The self-shrinking generator. In: De Santis, A. (ed.) Advances in Cryptology – EUROCRYPT 1994. Lecture Notes in Computer Science, vol. 950, pp. 205–214. Springer, Berlin, Heidelberg (1995). https://doi.org/10.1007/BFb0053436

68. Meliá-Seguí, J., García-Alfaro, J., Herrera-Joancomartí, J.: A practical implementation attack on weak pseudorandom number generator designs for EPC Gen2 Tags. Wirel. Pers. Commun. **59**(1), 27–42 (2011). https://doi.org/10.1007/s11277-010-0187-1

69. Meliá-Seguí, J., García-Alfaro, J., Herrera-Joancomartí, J.: J3Gen: a PRNG for low-cost passive RFID. Sensors **13**(3), 3816–3830 (2013). https://doi.org/10.3390/s130303816

70. Menezes, A.J., van Oorschot, P.C., Vanstone, S.A.: Handbook of Applied Cryptography. CRC Press, Boca Raton (1996)

71. Mihaljević, M.: A faster cryptanalysis of the self-shrinking generator. In: Pieprzyk, J., Seberry, J. (eds.) Information Security and Privacy – ACISP'96. Lecture Notes in Computer Science, vol. 1172, pp. 182–189. Springer, Berlin, Heidelberg (1996). https://doi.org/10.1007/BFb0023298

72. Mihaljević, M.J.: An approach to the initial state reconstruction of a clock-controlled shift register based on a novel distance measure. In: Seberry, J., Zheng, Y. (eds.) Advances in Cryptology – AUSCRYPT '92. Lecture Notes in Computer Science, vol. 718, pp. 349–356. Springer, Berlin, Heidelberg (1993). https://doi.org/10.1007/3-540-57220-1_74

73. Mihaljević, M., Zheng, Y., Imai, H.: A fast and secure stream cipher based on cellular automata over GF(q). In: Proceedings of the Global Telecommunications Conference, GLOBECOM 1998, vol. 6, pp. 3250–3255 (1998). https://doi.org/10.1109/GLOCOM.1998.775806

74. Mita, R., Palumbo, G., Pennisi, S., Poli, M.: Pseudorandom bit generator based on dynamic linear feedback topology. Electron. Lett. **38**(19), 1097–1098 (2002). https://doi.org/10.1049/el:20020750

75. Molland, H.: Improved linear consistency attack on irregular clocked keystream generators. In: Roy, B., Meier, W. (eds.) Fast Software Encryption-FSE'2004. Lecture Notes in Computer Science, vol. 3017, pp. 109–126. Springer, Berlin, Heidelberg (2004). https://doi.org/10.1007/978-3-540-25937-4_8

76. National Institute of Standards and Technology: A Statistical Test Suite for Random and Pseudorandom Number Generators for Cryptographic Applications, Special Publication SP 800–22 Rev. 1a (2010). http://csrc.nist.gov/publications/nistpubs/800-22-rev1a/SP800-22rev1a.pdf

77. National Tech. Info. Service: Data Encryption Standard. FIPS PUB 46 (1977)

78. Paar, C., Pelzl, J.: Understanding Cryptography. Springer, Berlin (2010)

79. Paterson, K.G.: Root counting, the DFT and the linear complexity of nonlinear filtering. Des. Codes Cryptogr. **14**(3), 247–259 (1998). https://doi.org/10.1023/A:1008256920596

80. Paul, G., Maitra, S.: RC4 Stream Cipher and its Variants. CRC Press, Taylor and Francis Group, Boca Raton (2012)

81. Pazo, M.E., Fúster-Sabater, A.: Cryptanalytic attack on the self-shrinking sequence generator. In: Dobnikar, A., Lotric, U., Ster, B. (eds.) Proceedings of the International Conference on Adaptive and Natural Computing Algorithms, ICANNGA 2011. Lecture Notes in Computer Science, vol. 6594, pp. 285–294. Springer, Berlin, Heidelberg (2011). https://doi.org/10.1007/978-3-642-20267-4_30

82. Peinado, A., Fúster-Sabater, A.: Generation of pseudorandom binary sequences by means of LFSRs with dynamic feedback. Math. Comput. Modell. **57**(11–12), 2596–2604 (2013). https://doi.org/10.1016/j.mcm.2011.07.023

83. Peinado, A., Munilla, J., Fúster-Sabater, A.: EPCGen2 Pseudorandom Number Generators: Analysis of J3Gen. Sensors **14**(4), 6500–6515 (2014). https://doi.org/10.3390/s140406500

84. Peinado, A., Munilla, J., Fúster-Sabater, A.: Improving the period and linear span of the sequences generated by DLFSRs. In: de la Puerta, J.G., Ferreira, I.G., Bringas, P.G., Klett, F., Abraham, A., de Carvalho, A.C., Herrero, Á., Baruque, B., Quintián, H., Corchado, E. (eds.) International Joint Conference SOCO'14-CISIS'14-ICEUTE'14. Advances in Intelligent Systems and Computing, vol. 299, pp. 397–406. Springer International Publishing, Cham (2014). https://doi.org/10.1007/978-3-319-07995-0_39

85. Petrovic, S., Fúster-Sabater, A.: Cryptanalysis of the A5/2 algorithm. IACR Cryptology ePrint Archive **2000**, 52 (2000). http://eprint.iacr.org/2000/052

86. Pries, W., Thanailakis, A., Card, H.C.: Group properties of cellular automata and VLSI applications. IEEE Trans. Comput. **C-35**(12), 1013–1024 (1986). https://doi.org/10.1109/TC.1986.1676709
87. Robshaw, M., Billiet, O. (eds.): New Stream Cipher Designs: The eSTREAM Finalists. Lecture Notes in Computer Science, vol. 4986. Springer, Berlin (2008). https://doi.org/10.1007/978-3-540-68351-3
88. Rueppel, R.A.: Analysis and Design of Stream Ciphers. Springer Verlag, New York (1986). https://doi.org/10.1007/978-3-642-82865-2
89. Rueppel, R.A., Staffelbach, O.J.: Products of linear recurring sequences with maximum complexity. IEEE Trans. Inf. Theory **33**(1), 124–131 (1987). https://doi.org/10.1109/TIT.1987.1057268
90. Simpson, L.R., Golić, J.D., Dawson, E.: A probabilistic correlation attack on the shrinking generator. In: Boyd, C., Dawson, E. (eds.) ACISP '98 – Third Australasian Conference on Information Security and Privacy. Lecture Notes in Computer Science, vol. 1438, pp. 147–158. Springer, Berlin, Heidelberg (1998). https://doi.org/10.1007/BFb0053729
91. Simpson, L.R., Dawson, E., Golic, J.D., Millan, W.L.: Lili keystream generator. In: Stinson, D.R., Tavares, S.E. (eds.) Selected Areas in Cryptography (SAC 2000). Lecture Notes in Computer Science, vol. 2012, pp. 248–261. Springer, Berlin, Heidelberg (2001). https://doi.org/10.1007/3-540-44983-3_18
92. Stembera, P., Novotný, M.: Breaking Hitag2 with reconfigurable hardware. In: 14th Euromicro Conference on Digital System Design, DSD 2011, pp. 558–563. IEEE Computer Society, Washington (2011). https://doi.org/10.1109/DSD.2011.77
93. Sun, S., Hu, L., Xie, Y., Zeng, X.: Cube cryptanalysis of Hitag2 stream cipher. In: Lin, D., Tsudik, G., Wang, X. (eds.) Cryptology and Network Security: 10th International Conference CANS 2011. Lecture Notes in Computer Science, vol. 7092, pp. 15–25. Springer, Berlin, Heidelberg (2011). https://doi.org/10.1007/978-3-642-25513-7_3
94. The ECRYPT Stream Cipher Project official website. http://www.ecrypt.eu.org/stream/
95. The eSCARGOT Project: European Stream Ciphers. Dept. of Electronic and Electrical Engineering. University of Sheffield. https://www.sheffield.ac.uk/eee/escargot
96. The original eSTREAM Project website. http://www.ecrypt.eu.org/stream/project.html
97. Tsalides, P.: Cellular automata-based built-in self-test structures for VLSI systems. Electron. Lett. **26**(17), 1350–1352 (1990). https://doi.org/10.1049/el:19900869
98. Verdult, R.: The (in)security of proprietary cryptography. Ph.D. thesis, Radboud University Nijmegen (2015)
99. Verdult, R., García, F.D., Balasch, J.: Gone in 360 seconds: Hijacking with Hitag2. In: 21st USENIX Security Symposium (USENIX Security 2012), Bellevue, WA, pp. 237–252 (2012). https://www.usenix.org/system/files/conference/usenixsecurity12/sec12-final95.pdf
100. Wolfram, S.: Cellular automata as models of complexity. Nature **311**, 419–424 (1984). https://doi.org/10.1038/311419a0
101. Wolfram, S.: Cryptography with cellular automata. In: Williams, H. (ed.) Advances in Cryptology – CRYPTO '85. Lecture Notes in Computer Science, vol. 218, pp. 429–432. Springer, Berlin, Heidelberg (1986). https://doi.org/10.1007/3-540-39799-X_32
102. Wolfram, S.: Theory and Applications of Cellular Automata (Including Selected Papers 1983–1986). In: Wolfram, S.E. (ed.) Advanced Series on Complex Systems, vol. 1, pp. 485–557. World Scientific Publishing, Singapore (1986)
103. Wolfram, S.: A New Kind of Science. Wolfram Media Inc., Champaign (2002)
104. Zeng, K., Yang, C.H., Rao, T.R.N.: On the linear consistency test (LCT) in cryptanalysis with applications. In: : Brassard, G. (ed.) Advances in Cryptology – CRYPTO '89. Lecture Notes in Computer Science, vol. 435, pp. 164–174. Springer, New York (1990). https://doi.org/10.1007/0-387-34805-0_16
105. Zhang, B., Feng, D.: New guess-and-determine attack on the self-shrinking generator. In: Lai, X., Chen, K. (eds.) Advances in Cryptology– ASIACRYPT 2006. Lecture Notes in Computer Science, vol. 4284, pp. 54–68. Springer, Berlin, Heidelberg (2006). https://doi.org/10.1007/11935230_4

106. Zhang, B., Wu, H., Feng, D., Bao, F.: A fast correlation attack on the shrinking generator. In: Menezes, A.J. (ed.) Topics in Cryptology – CT-RSA 2005. Lecture Notes in Computer Science, vol. 537, pp. 72–86. Springer, Berlin (2005)
107. Zhang, Y., Lei, J.G., Zhang, S.P.: A new family of almost difference sets and some necessary conditions. IEEE Trans. Inf. Theory **52**(5), 2052–2061 (2006). https://doi.org/10.1109/TIT.2006.872969

Printed in the United States
By Bookmasters